Disaggregating China, Inc.

A volume in the series

Cornell Studies in Political Economy
Edited by Peter J. Katzenstein

A list of titles in this series is available at www.cornellpress.cornell.edu

Disaggregating China, Inc.

State Strategies in the Liberal Economic Order

Yeling Tan

Cornell University Press
Ithaca and London

First published 2021 by Cornell University Press

Library of Congress Cataloging-in-Publication Data

Names: Tan, Yeling, author.
Title: Disaggregating China, Inc. : state strategies in the liberal economic order / Yeling Tan.
Description: Ithaca [New York] : Cornell University Press, 2021. | Series: Cornell studies in political economy | Includes bibliographical references and index.
Identifiers: LCCN 2020056483 (print) | LCCN 2020056484 (ebook) | ISBN 9781501759635 (hardcover) | ISBN 9781501759642 (ebook) | ISBN 9781501759659 (pdf)
Subjects: LCSH: World Trade Organization—China. | China—Economic policy—2000– | China—Economic policy—Political aspects. | China—Economic conditions—2000–
Classification: LCC HC427.95 .T36273 2021 (print) | LCC HC427.95 (ebook) | DDC 382/.920951—dc23
LC record available at https://lccn.loc.gov/2020056483
LC ebook record available at https://lccn.loc.gov/2020056484

For my parents

Contents

Figures and Tables

Tables

Acknowledgments

I owe the greatest debt to Peter A. Hall, whose counsel and scholarship has shaped incalculable aspects of this book as well as my approach to the study of politics. Over countless hours of conversation/interrogation, Peter posed difficult questions and tore apart my arguments while at the same time offering paths forward, channeling my focus toward a better iteration of a partially formed idea. More than these specifics, I have learned from him how to be tireless in the pursuit of a research question, to be exacting in formulating the answer, and to be a supportive and generous colleague. I hope that over the course of my career as I strive toward Peter's model of scholarly inquiry, I will be able to do justice to his teaching and mentorship.

Tony Saich has been a source of invaluable guidance since my days as a master's student, offering a broader perspective and wisdom at moments of doubt. He has pushed me to think about and analyze Chinese politics with greater nuance and to never take any one point of view at face value. Beth Simmons provided unstinting support and encouragement throughout every step of my doctoral student career, from the first international relations seminar to the anxieties of the job market and beyond. Her frank and precise comments pinpointed areas of weakness but in ways that always left me energized and motivated to do better. I benefited immeasurably from my conversations with Robert Z. Lawrence, many of whose insights into the political economy of trade and globalization have found their way in some form or another into this book. Together, these scholars formed a dissertation committee that a graduate student could only dream of having.

My gratitude extends to a broader community of scholars from whom I have drawn inspiration and whose advice and insights have helped me to become a better academic. Dani Rodrik sparked my interest in global economic governance and industrial policy and has provided encouragement and constructive critique in equal measure. Tom Christensen, Christina Davis, Will Norris,

Margaret Pearson, Ben Ross Schneider, Dan Slater, Rory Truex, Arne Westad, and the two anonymous reviewers from Cornell were incredibly generous in taking the time to read and offer comments on my draft manuscript. This book is much transformed from its early guise as a dissertation as a result of their sharp advice. I have also benefited from conversations with Suzanne Berger, John Donaldson, Henry Farrell, Archon Fung, Steph Haggard, Iain Johnston, Abraham Newman, Craig Parsons, Elizabeth Perry, Meg Rithmire, Susan Shirk, Ryan Sheely, Yuhua Wang, Michael Walton, and Mark Wu.

This research would not have been possible without the support of multiple institutions. My gratitude goes to the Harvard Weatherhead Center for International Affairs (WCFIA), the Fairbank Center's Desmond and Whitney Shum Fellowship, Harvard University's Frederick Sheldon Traveling Fellowship, and the China Scholarship Council for providing generous funding that made this book possible. My thanks go to Michelle Lamont, Erez Manela, Ted Gilman, and Clare Putnam, for creating a warm and stimulating environment in the WCFIA Graduate Student Associates program. The Ash Center for Democratic Governance and Innovation welcomed me to its events and provided several opportunities for me to present my work, and for this I am particularly grateful to Tony Saich and Kaori Urayama. Additional thanks go to the postdoctoral fellows at the Ash Center—Kyle Jaros, Sara Newland, Chen Huirong, Manfred Elfstrom, and Li Yao—for their feedback on various stages of my work. The China and the World program at Princeton and Harvard provided an ideal postdoctoral environment in which to turn this dissertation into a book, and my heartfelt thanks go to Tom Christensen and Iain Johnston for inviting me into the exceptional community of scholars that they have built over the years. Peter Katzenstein, Roger Haydon and Emily Andrew at Cornell University Press provided expert advice as well as good humor in shepherding this manuscript into its final form, and I am truly grateful for their guidance. I am further grateful to the University of Oregon's department of political science and the Oregon Humanities Center for their valuable financial support toward the publication of this book.

I am equally indebted to the institutions and scholars in China who made my fieldwork possible. I am grateful to Peking University's National School of Development and to Dean Yao Yang for welcoming me to Langrun Yuan and for providing a wonderful intellectual home while in China. I am also grateful for the warm collegiality extended by the Peking University faculty, including Professors Yu Miaojie, Yang Rudai, Xi Tianyang, Li Lixing, Zhao Bo, and Fu Caihui and many others. My thanks also go to the students from Peking University and the China University of Politics and Law for their superb research support. I am additionally grateful to Tsinghua University's School of Public Policy and Management for generously providing a stimulating professional environment during my multiple visits to China. My knowledge and understanding of Chinese economic policymaking have been greatly enhanced by conversations with Professors Yu Qiao, Hu Angang, Gao Yuning, Chen Ling, and Wei Xing. I would also like to extend my appreciation to the American Chamber of Commerce for generously sharing data from its annual Business Climate Survey. Additional thanks go to the International Center for Trade and Sustainable Development for kindly inviting me to its events in Beijing.

While many of them cannot be named, I would like to express my gratitude to the many interviewees I spoke with across various parts of China, the United States, and Geneva for generously taking time out of their busy schedules to share their thoughts with me. My knowledge and understanding of China and the World Trade Organization have been immensely enriched through these discussions.

A huge thank-you goes to my friends, who provided much-needed camaraderie, laughter, and perspective throughout what would otherwise have been a lonely and isolated endeavor: Hunt Allcott, Toby Carroll, J. P. Chauvin, Laura Chirot, Chua Ee-Lin, Edward Chew, Nilesh Fernando, Gwenda Fong, Chase Foster, Tomoko Harigaya, Alicia Harley, Tsin Yen Koh, Wendy Leutert, Elizabeth Linos, Niharika Singh, Russ Neu, Melissa Ong, Alec Worsnop, Catherine Z. Worsnop, Wu Wei Neng, and Xia Di as well as the many graduate students (from Harvard and beyond) I have met and overlapped with while doing fieldwork in China. Final thanks go to my family for unconditionally supporting my various career choices over the years and to James Conran for believing in me, for the emotional and intellectual sustenance that made this project possible, and for being a true partner in all of life's adventures.

Abbreviations

CCP	Chinese Communist Party
CPTPP	Comprehensive and Progressive Agreement for Trans-Pacific Partnership
EC	European Commission
FDI	foreign direct investment
GATS	General Agreement on Trade in Services
GATT	General Agreement on Tariffs and Trade
MIIT	Ministry of Industry and Information Technology
MOFCOM	Ministry of Commerce
MOFTEC	Ministry of Foreign Trade and Economic Cooperation
NAFTA	North American Free Trade Agreement
NDRC	National Development and Reform Commission
PBSC	Politburo Standing Committee
PPSC	provincial party standing committee
SAIC	State Administration for Industry and Commerce
SASAC	State-owned Assets Supervision and Administration Commission
SAQSIQ	State Administration of Quality Supervision, Inspection and Quarantine
SCM	Subsidies and Countervailing Measures
SDPC	State Development and Planning Commission
SEC	State Economic Commission
SETC	State Economic and Trade Commission
SOE	state-owned enterprise
SPC	State Planning Commission
SPS	sanitary and phytosanitary measures
TBT	technical barriers to trade
TPP	Trans-Pacific Partnership

TRIMS	trade-related investment measures
TRIPS	trade-related aspects of intellectual property rights
USTR	United States Trade Representative
WFOE	wholly foreign-owned enterprise
WTO	World Trade Organization

Disaggregating China, Inc.

Introduction

Integrating China into the Liberal International Order

In December 2001 China became the 143rd member of the World Trade Organization (WTO), successfully capping off a protracted fifteen-year effort to rejoin the international trading body. This event was arguably one of the most consequential turning points in the evolution of the global economic order and a significant milestone in China's own reform process. In 2000 China's share of world exports stood at 3.9 percent, far below that of leading exporters such as the United States (12.5%), Germany (8.6%) and Japan (7.6%).[1] Yet the international community anticipated that China would come to be a systemically important player in the global economy. Economically, the sheer potential size of China's domestic market and its growing exporting capabilities meant that its prospective membership in the WTO invited intense scrutiny and commercial demands from its trading partners. Politically, major global powers saw WTO membership as a means to integrate China's communist regime and command economy into the rules of the postwar international order.

US president Bill Clinton made China's WTO entry a political priority during his administration, arguing that WTO membership would "commit China to play by the rules of the international trading system, and bring China fully into that system in a way that will bring greater opportunity for its citizens and its industries."[2] At another event, Clinton claimed that it was "the most significant opportunity that we have had to create positive change in China since the 1970's, when President Nixon first went there."[3] It was this promise of integration that led US trade negotiators to seek a set of terms for China's WTO accession that would be comprehensive and tightly binding.[4] As a result, China's entry into the WTO involved not just substantial tariff reductions but also an unprecedented set of binding commitments to rewrite domestic policies relating to the rule of law, trading and distribution rights, government-enterprise relations, regulatory transparency, and more.[5] These US ambitions were matched by bold

1

determination on the Chinese side. President Jiang Zemin and Premier Zhu Rongji saw WTO entry not just as an important symbolic affirmation of China's rightful place in a key international organization but also a means to spur China's economic development and drive forward contentious domestic reforms. As such, China's long-awaited accession to the WTO was marked by a series of commentaries and articles in official state media that signaled China's commitment to the sweeping internal and external changes that were to come. English-language newspapers noted that "China will play an active and constructive role in promoting the development of world trade, and in establishing and perfecting a multilateral trading system."[6] In Chinese-language pieces, state media emphasized that WTO membership would "expedite the process of China's reform and opening up."[7] One of the major priorities of the day would be the "cleaning up of laws, regulations and policies contradictory to the rules of the World Trade Organization" and building an "impartial, efficient judicial system," while another was to speed up the reform of inefficient state-owned enterprises to prepare for an era of heightened market competition.[8]

Given the external pressure from major economic powers such as the United States and the domestic resolve of China's own leaders, there was widespread public anticipation that WTO membership would trigger broad-based liberalization and a weakening of the state-owned sector within China. Expectations were that as the rule of law and market competition expanded, government intervention would be circumscribed, and China's economy would come to operate on rules shared by major market economies. China would go from "shallow" to "deep" integration with the rest of the world, and some even speculated that the economic liberalization would provide an enabling force for domestic political change.[9]

Almost two decades later the high hopes surrounding China's WTO accession have dissipated, replaced by pessimism and foreboding. In January 2018, United States Trade Representative (USTR) Robert Lighthizer declared that the United States had "erred in supporting China's entry into the WTO," effectively calling an end to a decades-long bipartisan policy of engaging China through multilateralism.[10] The USTR's annual report on China asserted that "WTO rules are not sufficient to constrain China's market-distorting behavior" and that hopes that WTO membership would alter China's statist economic policies had been "disappointed."[11] In essence, the report made the case that China's mode of economic governance was "incompatible with the open, market-oriented approach expressly envisioned by WTO members and contrary to the fundamental principles running throughout the many WTO agreements."[12] Strikingly, these views were not restricted to members of the Republican Donald Trump administration. Also in 2018, former senior Barack Obama administration officials Kurt Campbell and Ely Ratner argued that "the liberal international order has failed to lure or bind China as powerfully as expected," calling for a reassessment of the US policy toward China.[13] While those declarations of engagement failure were met with disagreement by other former senior government officials from both parties, the shift in policy discourse from consensus to dissensus nevertheless reflected the emergence of major tensions within the US government over its China policy.[14] Nor was the United States alone among major developed economies in expressing qualms over the compatibility of China's economic system with WTO rules. The European Union, along with the United States, has opposed having China be given status as a market economy under WTO rules on trade remedies on the basis that heavy state intervention continues to distort the economy.[15]

These negative sentiments in the United States soon manifested themselves in the launch of a sharply aggressive trade policy against China. In 2018, the US government in a series of announcements imposed tariffs on a range of Chinese exports potentially totaling up to $450 billion while also lodging a complaint with the WTO dispute resolution body over China's technology licensing arrangements.[16] China responded with its own set of retaliatory tariffs, and it was only in late 2019 that both sides came to an agreement on a "phase one deal" to temporarily suspend the tit-for-tat trade war.[17] In early 2020 trust between the United States and China dipped even further as countries around the world struggled to respond to the COVID-19 pandemic, and anger and suspicion rose against China over issues of transparency and disclosure in its handling of the initial outbreak.

On its part, while the Chinese government has continued to voice its support for the international trading system, it has yet to provide a clear articulation of the principles that will guide its international behavior in a way that provides assurance to its major trading partners.[18] Although China had been gradually shifting away from its stance of maintaining a low profile to adopt a more assertive foreign policy position, the signals sent by the Chinese leadership have so far been consistently opaque.[19] In a speech in Moscow in 2013, Chinese president Xi Jinping called for the international community to "build a new type of international relations with win-win cooperation at the core," and two years later in Seattle he repeated the phrase in an address to the National Committee on US-China Relations.[20] In 2016 in a speech commemorating the ninety-fifth anniversary of the founding of the Chinese Communist Party (CCP), Xi said that "the CCP and the people of China are fully confident that they can provide a *China solution* for humanity's search for better social institutions."[21]

These themes were emphasized again in 2017 at the CCP's Nineteenth Party Congress wherein Xi cemented his political position as the most powerful man to lead the country since Mao Zedong. Xi's report to the congress reiterated that China offers "a *Chinese approach to solving* the problems facing mankind" while also stressing that "socialism with Chinese characteristics has crossed the threshold into a *new era*" and that "China champions the development of a *community with a shared future for mankind,* and has encouraged the evolution of a global governance system."[22] However, the substantive meaning of phrases such as "new type of international relations" and the "China solution" has yet to be clarified. At a time when China's outbound foreign investment has been surging alongside China's creation of new institutions such as the Asian Infrastructure Investment Bank and the "Belt and Road Initiative" for expanding overseas infrastructure and investment projects, scrutiny over China's domestic and external economic policies has intensified. The lack of specificity in the Chinese leadership's pronouncements leaves the outside world to speculate about the implications of a more assertive China for the global economic order.

To what extent has WTO membership succeeded in integrating China into the international rules-based system? This question strikes at the heart of the liberal internationalist project. The core promise of liberal internationalism has been that through the principles of open and mutually beneficial economic exchange, rules-based engagement, and inclusivity, an expanding diversity of economies would be brought closer together under a common system of governance, in the process "undercutting illiberal tendencies and strengthening the fabric of international community."[23]

However, the degree to which these expectations would be fulfilled has never been certain. WTO entry introduces pressures on state officials to adjust to two new aspects of economic governance. The first involves policymaking under conditions of open economic competition, which creates not just economic winners and losers but also demands on state officials to rethink their policies in light of the new threats and opportunities brought by economic integration. In addition, as chapter 1 discusses, WTO rules impose a set of terms and conditions specifying the manner in which liberalization is to occur. This means new external obligations setting out the ways in which state authorities can and should govern a range of domestic domains, from health and safety standards to intellectual property protection and industrial promotion. In other words, WTO entry involves not just economic liberalization but also institutional restructuring, a process that potentially alters the distribution of power across different parts of the state and thereby directly implicates the interests of the bureaucracy.

The political demands of WTO entry on domestic policymaking was perhaps uniquely strenuous for China, given its authoritarian structure, size, legacy of central planning, and heterodox approach to reform. While WTO rules broadly reflect the principles of market liberalism, China's domestic economic governance at the point of accession was still strongly influenced by command economy practices and even after decades of reform continued to be characterized by policies that deviated from the tenets of open market competition. Moreover, the CCP continued to retain ownership of and control over major strategic portions of the economy. And crucially, although China's top leaders sought to use WTO membership to credibly commit to domestic liberalization, the depth and breadth of these commitments might generate resistance from parts of a bureaucracy that did not want its discretion circumscribed by external rules as well as opposition from powerful industry interests fearful of import competition.[24] Even at the broad societal level, the prospect of WTO entry stirred up intense public debate within China over whether or not liberalization would be of benefit to the nation.[25]

These issues, as chapter 1 details, are complicated by the enormity and complexity of China's domestic political structure. With a ruling party whose membership surpasses the population of Germany and with a governance structure that not only delegates many economic functions down to subnational levels but is also duplicative and overlapping between party and state at each level, China has a politics of economic governance that spans multiple dimensions simultaneously. This book therefore focuses on the state as a site of conflict, examining the ways in which the demands of WTO entry generated major tensions that intensified and transformed China's intrastate politics. The analysis gets at these politics by disaggregating "China, Incorporated." Rather than treat the state as a cohesive unitary actor, this book unpacks the impact of WTO entry on the incentive structure of disparate actors within the party-state and then elucidates the channels through which WTO membership shaped the various policy responses of these actors.

Given China's size, legacy of economic planning, and authoritarian regime, its WTO entry can be considered a hard case of the manner and extent to which international economic organizations can successfully integrate new members. It is also a critically important case. In a world economy that runs on interdependent globalized supply chains, shifts in China's domestic economic governance create ripple effects that are felt in other parts of the world. The stability and functioning of the global economy depend in part on the degree to which China abides by, undermines, adjusts, or overturns the existing rules of international economic institutions. Moreover, the stringent terms of China's WTO accession reflect a broader move over time toward trade agreements requiring deep terms of integration, containing ever more intrusive commitments aimed at altering domestic institutions. At the same time, a growing diversity of nation-states are not only joining the WTO but also participating in these ambitious agreements, making the scope of domestic adjustment even more daunting. The participation of Vietnam and Brunei in the Comprehensive and Progressive Agreement for Trans-Pacific Partnership is just one such example. Questions regarding the ways in which states such as China respond to membership in international institutions and why they adopt one set of responses over another hold theoretical significance and also bear substantive policy importance.

The Argument

What were China's policy responses to WTO entry, and what explains China's economic policy trajectories in the WTO era? This book seeks to do two things: first, to capture and establish the major economic policy trajectories that emerged from China's entry into the WTO, and second, to explain the pattern of these trajectories. I approach these tasks in a series of steps. The first is to conceptualize the policy options available to the Chinese government in responding to WTO entry in terms of three competing *state strategies*. I then disaggregate the "black box" of the Chinese state into its component leadership and governmental actors in charge of economic policymaking and lay out their differential political interests vis-à-vis WTO entry. To capture the enactment of these three state strategies by various substate actors over time, I collected an original data set of industry regulations issued by all the various arms of the Chinese state at different levels of authority. I then apply automated textual analysis to the corpus of documents, thereby uncovering different approaches to economic governance latent in the bureaucratic language deployed by Chinese substate actors. In so doing, I establish that different economic stakeholders within the state in fact responded to WTO entry with a variety of divergent policies. Finally, I explain the factors driving this policy divergence as being manifested in three important dimensions of China's economic structure: across administrative levels, within the central state between the political leadership and its economic bureaucracy, and across industries.

While existing literature more commonly frames responses to economic integration in the dichotomous terms of liberalization versus protectionism or compliance versus defiance, this study takes the view that actors within the state in fact can choose from a broader range of policy responses. As WTO entry introduces not just a more openly competitive economic environment but also additional terms and conditions for policymaking, in crafting their responses governmental actors have to consider not simply whether to intervene in the economy but also the specific manner in which to do so. I propose to characterize these responses as state strategies, where a strategy is defined as *a set of policy instruments, rooted in a particular mode of state-market relations, that is adopted to further a state's economic objectives.*

Contestation over how to govern the economy exists in all societies, even in a single-party regime such as China. While planned economy approaches dominated during the Maoist years, the range of strategies available to actors within the state has expanded through an incremental and uneven process of reform. Experiments to adopt market-based mechanisms—such as price reform—were enacted on a dual-track basis, allowing market forces to grow around and alongside, rather than replacing, existing command-based instruments.[26] As chapter 1 documents, these types of reforms expanded at an uneven pace, some scaling up to be adopted nationwide and others remaining circumscribed. Therefore, even at the eve of China's WTO entry, contrasting governance approaches coexisted across different parts of the state in a heterogenous manner, reflecting internal disagreements over the appropriate manner in which to govern the economy.[27]

Suppose that in November 2001, the month before China's WTO accession, the fictional city of Xintao wanted to expand its microwave industry. What policy options would be available to the Xintao government? I propose that government actors in China can choose from among three distinct state strategies in their economic policymaking: market-substituting (directive), market-shaping (developmental), and market-enhancing (regulatory). Although these strategies are all oriented toward economic growth, they differ in the state-market relations undergirding each approach.

First, the Xintao city government could rely on long-standing administrative measures available to it from the days of the planned economy and deploy a *directive* or market-substituting strategy for growth. Rather than letting the market (meaning independent firms responding to price signals) determine the outcome, Xintao authorities could issue administrative directives that take the place of market functions. By setting targets for the number of microwaves to be produced and the number of firms to engage in production, the government would shut down small and inefficient firms and force enterprise mergers to create larger-scale production units, thereby directly determining the structure, composition, and output of the microwave industry.

Alternatively, the Xintao authorities could take a more *developmental* or market-shaping approach inspired by Japanese or Korean industrial success and seek to promote firm activity in the microwave sector through targeted incentives. For example, city leadership could issue tax breaks and subsidies to attract investment from technologically advanced foreign microwave companies and encourage partnerships with local firms. In contrast with the directive strategy, the behavior of firms is not dictated by administrative measures. Rather, the Xintao city government's developmental strategies aim to influence firm decisions via targeted incentives.

Finally, Xintao city could adopt a *regulatory* or market-enhancing approach more commonly found in developed market economies and advocated by WTO principles. In this approach, Xintao authorities would try to promote industrial growth in the microwave sector by ensuring that institutions required for the market to function properly are in place, from property rights and the rule of law to setting clear product and consumer safety standards. All firms, regardless of whether they are domestic or foreign, large or small, would then compete on a level playing field, and the number of microwaves ultimately produced and sold would be determined by market competition.

In reality, examples of all three strategies can be found in Chinese economic policies, even within a single industry. The metal products industry, for example, is governed by measures that reflect a mixture of directive, developmental, and regulatory approaches. In 2012, the Liaoning provincial government issued a policy announcing a four-year plan to rectify and restructure the metal and nonmetal mining industry in the province. The policy emphasized shutting down mines whose scale was too small or that used backdated technology or other practices that threatened to hold back industry development. Specific targets and timelines were established with the goal of achieving economies of scale and industrial consolidation. Within four years, a total of

471 mines would be shut down in accordance with a predetermined schedule: 94 mines in 2012, 129 mines in 2013, and so on. These measures reflected classically directive approaches relying on government diktat to reshape industry outcomes.[28]

Yet in 2008, the Hechi city government in Guangxi province issued a policy aimed at kick-starting indigenous innovation in the nonferrous metal processing industry. This policy document approached the goal of industrial development by emphasizing technology upgrading through cooperation with tertiary and scientific research institutions and the expansion of industrial linkages. Rather than direct intervention, the policy focused on attracting skilled labor into R&D-intensive segments of the industry by offering stock options and higher salaries as well as incentives to encourage high-tech firms to form industry clusters in special industrial parks—all market-shaping policies designed to attract firm activity and human capital into a specific industry.[29]

That same year, in contrast, the Shanghai city government issued a regulation aimed at strengthening arms-length supervision of the metal products industry. The notice issued by the Shanghai City Quality and Technical Supervision Bureau accredited the Shanghai Metal Materials Quality Supervision and Inspection Station with certification authority over a range of metal products, citing the Product Quality Law, the Standardization Law, and the Metrology Law. Unlike the Liaoning notice that directly intervened to achieve industrial consolidation and the Hechi policy aimed at incentivizing technology upgrading, Shanghai's notice emphasized the existing body of regulations and laws surrounding product quality and standardization. Industry outcomes would depend on the ways in which firms operated within these regulatory guidelines rather than through direct targets or indirect incentives.[30]

Why is it that such vastly different approaches to economic governance can be found in the industrial policies of the Chinese government, even within a single industry? This is in part a product of China's incremental and experimental reform process and also because authority over economic development is decentralized within the party-state, allowing different governmental actors substantial policy autonomy. At the same time, this authority is a highly valued asset that affects prospects for political advancement within the CCP system. Therefore, economic policies are not issued lightly. Government actors adopt different policy measures with an eye toward advancing their position within the bureaucracy and are likely to approach their responses to WTO entry with a strategic calculus rather than simply surrendering their policy authority to an external set of rules or ceding to the demands of domestic interest groups.

China's policy responses to WTO entry therefore will be mediated by a domestic governance structure that, while authoritarian, is highly decentralized and riven by internal political competition. As authority over economic policymaking is distributed across multiple parts of the state, this book casts a wide net to examine policy responses across three dimensions of analysis, all of which are consequential to the political economy of China. The first dimension, which is the subject of chapter 3, is concerned with varying policy responses across China's administrative hierarchy: the central, provincial, and subprovincial (local) governments that make up China's quasi-federal system. This dimension is highly significant given the devolution of economic policy to subnational levels and the disconnect between central-level priorities and political-economic concerns at lower-level jurisdictions. Second, chapter 4 examines policy responses within the central party-state apparatus, where multiple interests compete over the formulation of economic policy. This disaggregation is of analytical value because policy conflict takes place between not only powerful central agencies with opposing approaches to governance but also the party leadership and the state bureaucracy. Therefore, policy trajectories that appear to be emanating from Beijing can in fact be driven by distinct interests within the vast central party-state complex. Chapter 5 turns to the consequences of center-subnational policy conflict for governance across "national champion" industries in China's WTO era, a topic of core relevance to the relationship between WTO entry and economic policymaking. The value in undertaking this multiactor, multidimensional analysis is that it allows us to avoid the trap of drawing generalized conclusions about the behavior of an entity as large and complex as the Chinese state from observing just one aspect of its governing structure. Instead, this book offers a framework that captures the multifaceted ways in which China's domestic political structure mediates responses to the economic and bureaucratic demands of WTO entry.

Although the analytical breadth of this study is wide and cross-cutting, my framework for explaining the policy trajectories emanating from different parts of the state is a simple one. I show that in broad terms, the strategic responses adopted by different government actors to WTO entry is determined by their varying *probability of being sanctioned* and *prospects for political advancement*. With this straightforward cost-benefit approach, however, this book stresses—and demonstrates—that the forces of sanction and advancement work through a variety of channels inside China's complex domestic structure. The relevant combination of sanction and advancement that drives responses to external integration therefore shifts depending on the set of actors being observed.

WTO membership introduced a new external source of discipline on Chinese state behavior that manifested itself in several forms. As a newcomer to the trading system, China was under broad reputational pressure to enact policies consistent with WTO rules.[31] This pressure was strengthened through a transitional review mechanism allowing other WTO members to regularly monitor China's implementation of its commitments. Additionally, the WTO dispute settlement mechanism imposed enforcement pressures by raising the threat of retaliation should one fail to abide by dispute panel rulings. The degree to which this external pressure is felt by different actors within the party-state, however, varies depending on domestic channels of enforcement. As noted earlier, these enforcement channels are often weak due to China's internally fragmented governance structure. The probability that a given substate actor might be punished for violating China's WTO obligations therefore varies depending on where that actor sits in the governing structure and on the nature of its relationship with higher-level authorities.

WTO entry further has the potential to shift a given governmental actor's prospects for political advancement, defined broadly as the attainment of greater rank or influence within the state. Political competition within the CCP takes place through formal and informal channels. Formally, one path through which CCP members work their way up the party hierarchy is the cadre evaluation system, where promotion prospects depend to a large degree on delivering economic growth.[32] WTO entry affects these promotion prospects by introducing new economic opportunities (foreign capital and export markets) as well as threats (import competition and unemployment) that might favor the growth prospects of some jurisdictions and industries but hurt that of others. At elite levels of the party, however, political competition takes place more informally, between competing party factions as well as between economic agencies angling for greater control and influence over the economic agenda.[33] With WTO entry, some central government actors might find that international trade rules potentially undermine their influence over economic policy, while others see new opportunities to deploy China's WTO commitments as leverage to advance their bureaucratic position.

In other words, despite single-party rule, China's policy responses to WTO entry depend on much more than just the preferences of the top leadership. It might be tempting to reason that compared to democratic regimes, an authoritarian state such as China—one that has proved resilient through periods of intense domestic political turmoil such as the Great Leap famine, the Cultural Revolution, and the Tiananmen Square crackdown and has demonstrated the state capacity to produce an unprecedented four decades of nearly continuous economic growth—would have a strongly coherent response to WTO entry; that is, China would either comply or violate its WTO obligations depending on the intentions of the party leadership. This book shows, however, that there was no monolithic state response, be it compliance rather than defiance or liberalization instead of protectionism, and that these dichotomous categories are inadequate in capturing the meaningful dimensions of a state's responses to international integration. Rather, because China's domestic governing structure is simultaneously fragmented and highly competitive, WTO entry provokes divergent responses (directive, developmental, and regulatory) from different actors within the state and over time, driven in broad terms by the varying probabilities of being sanctioned and prospects of advancement.

Across administrative levels, decentralization of responsibility over economic policymaking means that while subnational states hold substantial authority over their economic jurisdiction, they are simultaneously sheltered from the pressures of international economic diplomacy and from the threat of direct punishments from violating WTO rules. If the Hebei provincial government undertakes a large promotional push to develop the solar panel industry, for example, which leads to a surge of low-priced solar panels on world markets, Hebei's actions might provoke a WTO case claiming that China's solar industry measures are WTO-inconsistent. However, the reputational and litigation costs associated with managing this dispute are largely borne by the central government that is entrusted with the sovereign duties of representing China at the international level as well as enforcing panel rulings domestically. Therefore, to the extent that the threat of external sanction shifts the state toward a more regulatory economic strategy for growth, this threat is strongest at the central level of government and weakens progressively at subnational provincial and local levels.

At the same time, while government leaders at lower administrative levels are somewhat insulated from the threat of external sanction, their advancement prospects are directly implicated by the open economic competition introduced by WTO entry. These new market conditions can either enhance or threaten their economic performance, thereby profoundly affecting their promotion prospects. In this regard, the industrial diversification of a jurisdiction plays an important role in affecting whether a government at a given administrative level stands to benefit or lose from economic liberalization. Provinces, being many times larger than local units such as counties and cities, are likely on average to be more highly diversified. Consequently, WTO entry is more likely to bring stronger export opportunities relative to import competition for the average provincial authority, driving it to respond with broadly developmental strategies to take advantage of new export markets. In contrast, the average local government oversees a more poorly diversified jurisdiction that is more vulnerable to the risks of import competition, as any given industry that suffers a negative import shock is likely to occupy a larger share of that jurisdiction's economy. The average local authority is hence more likely to adopt directive measures that forcefully intervene in mitigating the dislocations brought by WTO entry.

The strategic responses emerging from Chinese governments at varying administrative levels are therefore shaped concurrently by these two factors: the threat of sanction for violating WTO rules, which weakens at lower levels of the state, and the ability to improve one's promotion prospects by ensuring economic growth under conditions of open competition, which also becomes more challenging at lower (less diversified) administrative levels. As chapter 3 shows, the central government, being most exposed to the threat of WTO sanction relative to subnational units, adopted a largely regulatory response to WTO entry. The provincial government, being one level removed from WTO sanction yet large enough to oversee a fairly diversified industrial base, adopted a largely developmental response to the same event. The local state, being most sheltered from international rules and yet overseeing on average a poorly diversified industrial base, shifted toward directive strategies aimed at mitigating the threat of import competition. The impact of WTO entry on Chinese economic governance therefore varied systematically within China's hierarchical state, triggering divergent strategic responses from different government authorities rather than a convergence toward liberalization.

This analysis partially corroborates initial assessments of Chinese economic policies in the early 2000s that pointed to the strengthening of regulatory market-enhancing institutions.[34] What I show additionally is that this turn toward regulatory modes of governance was predominantly driven by the central government, while subnational levels adopted starkly divergent strategies. By the second half of the decade, however, scholars were increasingly calling attention to a more active industrial policy as well as the growing assertiveness of a state capitalism associated with the consolidation of central state-owned enterprises and heightened central state intervention in strategic industries.[35] What lies behind this reversal in central policy trajectory? Chapter 4 turns to explaining the rise of Chinese state capitalism by looking within the central government to ask why the establishment of market-enhancing institutions in response to WTO entry did not constrain the subsequent strengthening of developmental policies. What explains the timing of this policy change?

To understand these shifts in the central party-state's policy trajectories, I disaggregate the powerful Beijing government into its component actors who wield substantial authority over the direction of economic policymaking, namely the party leadership and central bureaucratic agencies. While conventional explanations for the rise of Chinese state capitalism point either to the 2008 global financial crisis or a change in preferences associated with new leaders Hu Jintao and Wen Jiabao, I show that the developmental turn occurred around 2006 and therefore preceded the financial crisis, while preferences on reform did not differ dramatically between leaderships. Instead, the shift in policy trajectory—and the timing of that shift—rests on the ways in which WTO rules interacted with the operation of sanction and advancement prospects within the central party-state.

In terms of sanction, while party leaders in the WTO era were largely supportive of regulatory reforms, the central bureaucracy was split between agencies supporting regulatory and developmental approaches to governance. Therefore, shifts in policy trajectory cannot be satisfactorily attributed to a change in leadership preferences but instead rests on whether economic policies are being driven by the leadership or central agencies. Contrary to assumptions that authoritarian leaders are able to achieve their goals through top-down direction, this book emphasizes that the direction of central economic policies depends on the political relationship between the party leadership and its central bureaucracy as well as on the politics of reciprocity that run through leadership networks. Specifically, is the leadership able to sanction the bureaucracy and effectively deploy the state as its *agent* for implementing party priorities, or does it instead have to rely on the bureaucracy as a key *constituency* for political support? When sanctioning is effective, the political leadership is able to drive China's WTO response. In the latter scenario, policy responses are instead more directly shaped by the central bureaucracy. The resulting strategies then depend on the WTO's impact on the advancement prospects of different agencies. While agencies supportive of regulatory governance were able to use China's WTO commitments as external leverage to advance their bureaucratic influence, developmental agencies saw the same rules as imposing an external constraint on their policymaking.

In the years immediately following WTO entry, the Jiang-Zhu leadership was able to effectively discipline the central bureaucracy and successfully deployed the state as its agent to adopt regulatory reform. At the same time, proliberalization regulatory agencies used the external pressure to implement new WTO rules as leverage to advance their domestic influence relative to their developmental rivals. As a result, China's early WTO years were associated with a strong regulatory response from the central government. With the leadership changeover in 2003, however, the new leaders (Hu and Wen) were not able to effectively discipline the central bureaucracy and instead needed to rely on the state as a key political constituency. This shift in political relations then allowed the central bureaucracy greater influence over the trajectory of economic governance. At the same time, the degree to which regulatory agencies could use WTO rules as leverage to advance their influence weakened as the implementation schedule for China's WTO commitments was largely concluded, leaving these agencies in a weakened position relative to their developmental rivals in controlling the economic agenda.

This dilution of WTO leverage, combined with the diminished ability of the leadership to discipline its bureaucracy, allowed developmental agencies new room to strengthen their enactment of developmental policies, producing what has come to be characterized as the rise of "Chinese state capitalism." This shift in trajectory, as chapter 4 shows, was not driven by a coherent central strategy. Instead, the strengthening of regulatory policies in the wake of WTO entry was driven by a coalition of the Jiang-Zhu leadership and proreform agencies using the WTO as external leverage to drive policy change. As the strength of WTO leverage faded over time, this market-enhancing policy trajectory was subsequently superseded by more active industrial policies promoted by rival developmental agencies under a weakened Hu-Wen leadership.

With central and provincial governments now aligned in the deployment of developmental strategies, did industrial policy in China therefore become stronger and more coherent in the latter half of the 2000s? Chapter 5 examines policy trajectories across China's manufacturing industries and explains why, despite a turn toward developmentalism at both national and subnational levels, WTO entry nevertheless exacerbated policy conflict in industry governance through a foreign direct investment (FDI) channel.

As I have emphasized, poor monitoring and enforcement channels within the state means that the central government is often unable to sanction subnational governments for pursuing policies at odds with national priorities. This enforcement gap is consequential because although both central and subnational governments seek to enhance economic growth, they are motivated by different *types* of political advancement and therefore favor the enactment of developmental strategies in different industries. While subnational governments are driven to achieve rapid short-term growth results so as to enhance their prospects for *rank* advancement, the central government is concerned with technological upgrading and long-term drivers of growth that boost the CCP's prospects for overall *regime* advancement. Although this tension has been a constant feature of China's political economy, WTO-led liberalization enhances the access of subnational governments to external resources in the form of FDI, boosting the ability of these governments to enact policies that bypass central priorities. The central government prioritizes the deployment of developmental policies that strengthen the absorption of advanced technology embedded in FDI into "national champion" industries so as to improve the regime's long-term durability. In contrast, subnational governments emphasize developmental strategies that maximize the quantity of foreign capital inflows into industries with the greatest potential for short-term growth in the service of enhancing their rank promotion prospects. Consequently, FDI brings greater center-subnational conflict into industrial policies at precisely a time when liberalized competition is provoking the center to boost its promotion of national champion industries.

The core points of my argument are as follows. First, dichotomous frameworks are insufficient in capturing the full range of ways in which states can and do respond to membership in international economic institutions. Too often, analyses cast the state as moving in either a positive "reformist" direction or a negative "backsliding" direction that does not map onto the multiple potential strategic options that a state can adopt in its governance response. The common treatment of the state as a unitary actor in international affairs moreover overlooks the possibility for more than one response to emerge from an event such as WTO entry, given that the state comprises multiple actors whose interests are more often than not misaligned or even sitting in opposition to one another. This applies even in the particularly hard case of China, an authoritarian single-party regime that is often viewed as having strong state capacity and resilience.[36]

Indeed, this book stresses that authoritarian rule does not translate into more monolithic or top-down responses to international integration. Given China's size and fragmented governance structure, actors within the state can and do respond to the new conditions introduced by WTO liberalization with their own political logic, choosing from a range of state strategies for economic growth that are grounded in distinct state-market relations. Therefore, despite authoritarian rule, what we observe from China is multiple responses to WTO entry that shift over time whereby an actor's choice of strategy depends on how new trade rules and liberalized competition affect their probability of being sanctioned and prospects for political advancement. While the explanatory framework is simple, the forces of sanction and advancement manifest themselves in several channels within China's complex governing structure. Sanction can arise externally from WTO mechanisms or internally from the party leadership or the central government, while the probability of being disciplined through one of these channels depends on where one sits in the state apparatus. Advancement can mean rank advancement for cadres at lower administrative levels, greater bureaucratic influence for central agencies, or regime advancement at national levels, and WTO entry shifts an actor's prospects for each of these types of advancement differentially.

Therefore, while international trade rules are increasingly designed to bind members to a set of common principles, the resulting responses may not necessarily lead to a convergence in state behavior. These responses also do not simply reflect the poor policy implementation that characterizes much of Chinese governance, as much important scholarship on the political economy of China has emphasized. Instead, WTO entry affected the behavior of governmental actors differentially, provoking these actors to *intensify* their economic strategies but along divergent paths across different parts of the state and over time.

Research Methodology

One central challenge that this study confronts is the question of how to measure the outcome of interest, or the three state strategies (regulatory, developmental, and directive), across multiple dimensions of the state and over time. I approach this by employing quantitative textual analysis of an original data set of Chinese industry regulations to capture and assess variation in strategic responses. The data set was collected by searching for official government laws, regulations, and measures related to China's manufacturing sectors in a comprehensive online repository of Chinese laws and regulations and then scraping all of the search results from the website. The resulting corpus of documents comprises over forty thousand legal and administrative documents issued by diverse arms of the central government—the People's Supreme Court, the National People's Congress, the State Council, economic ministries, regulatory bodies, and so on as well as subnational authorities from provincial down to county governments—covering 128 manufacturing industries from 1978 (the start of China's reform period) to 2014.

What this corpus contains is not just a comprehensive collection of industrial policies and measures but also the *language* of the Chinese bureaucracy as employed by a multiplicity of actors—influential and weak, national and subnational, new and old—who are or have been involved in economic governance over the course of the post-Mao reform period. The documents therefore capture different approaches to the question of how—or in what ways—to grow the economy, which as chapter 1 discusses are deeply entrenched within the party-state and have been a constant feature of Chinese political economy even during periods of remarkable growth.

These contrasting approaches to governance show up in the deployment of different vocabulary in the industrial policies and measures issued by different government actors. I identify the vocabulary associated with the regulatory, developmental, and directive state strategies by using an unsupervised machine learning method in text analysis that allows us to uncover various "topics" (or clusters of words) that are latent within the corpus of Chinese industry documents.[37] By locating topics that are reflective of these three state strategies, I can then track the prevalence of each strategy in policy documents adopted by different arms of the Chinese party-state for different industries and over time. In the quantitative analyses, I additionally use data on China's industrial tariffs, imports and exports, state ownership, and employment as well as province-level economic and government data to test my explanations for why a particular state strategy is adopted over another in response to WTO entry.

These quantitative assessments are complemented by qualitative analysis using a range of primary and secondary material. Qualitative data pertaining to the two sets of general secretaries and premiers in the main periods of analysis prior to and after WTO entry (Jiang and Hu, Zhu and Wen) are drawn from their public speeches, public and internal records of leadership meetings with domestic and foreign officials and businesspeople, and internal government memos. Other sources include English- and Chinese-language media reports as well as articles written by government officials in journals published by party or semiofficial institutions. Additional qualitative evidence related to bureaucratic politics and the politics surrounding China's WTO accession was gathered through interviews with current and former officials in the Chinese and US governments and at the WTO as well as industry representatives, scholarly experts, and journalists.

I also explore the impact of WTO entry on industry governance through two case studies on the automobile and semiconductor industries. Empirical evidence for these case studies is drawn from official government speeches and policies, annual industry reports, media reports from general newspapers, specialized industry journals, and firm-level data. The analyses further draw on data collected in interviews in Beijing, Shanghai, Jiangsu, Guangdong, and Yunnan with Chinese and foreign industry managers, journalists on the business beat, industry association representatives, academics, and members of semiofficial think tanks such as the Chinese Academy of Science and the China Development Research Foundation.

In chapter 1, I explain why WTO rules have a profound impact on bureaucratic interests and why the terms of China's accession to the WTO was an important external shock to the Chinese bureaucracy. By highlighting the debates over the direction and goal of reform in the years leading up to WTO entry and why China's incremental and uneven process of reform has generated internally divergent approaches to governing the economy, I further underscore why any analysis of China's WTO policy responses needs to take into account internal disagreements over economic policy. Chapter 2 presents the theoretical framework of the book, beginning by outlining the three competing state strategies through which different government actors can respond to WTO entry. I then offer my explanation for why WTO membership provoked a divergent rather than monolithic response from these actors, highlighting the ways in which the varying likelihood of sanction and the prospects for political advancement combine to drive each set of actors to adopt different responses. Chapters 3 through 5, the main empirical chapters, explain the divergent strategic responses across three dimensions of economic governance: across administrative levels, between the CCP leadership and central economic agencies, and across different industries. The conclusion discusses the various contributions of this study to international relations and the political economy of China by highlighting the value of disaggregating state responses to global economic integration. The discussion then turns to the ways in which the findings of this book apply beyond the case of China in expanding our understanding of the relationship between global integration and the domestic politics of economic policymaking.

Chapter 1

China's WTO Challenge

The establishment of the World Trade Organization (WTO) in 1995 was a critical turning point in the evolution of the global trading regime, an event that substantially increased the scope and depth of multilateral trade cooperation. No longer would trade negotiations be restricted to the reciprocal lowering of border barriers such as tariffs and quotas, as was the objective of the General Agreement on Tariffs and Trade (GATT). Instead, attention expanded to the dismantling of nontariff barriers as avenues for liberalizing trade.[1] The cooperative principles established at the WTO, it was argued, represented a move toward a deeper version of liberal internationalism—a less hierarchical and more inclusive order based on universal and binding rules, with coverage over expansive domains.[2] This push toward inclusivity on the one hand and deep integration on the other generates tricky tensions for countries that have joined the WTO since 1995, as they increasingly represent a variety of regime types and stages of development: China in 2001, Vietnam in 2007, Russia in 2012, and Afghanistan in 2016.

The WTO's body of rules is articulated in a set of agreements, including GATT, the General Agreement on Trade in Services (GATS), and the agreements on sanitary and phytosanitary measures (SPS), subsidies and countervailing measures (SCM), trade-related investment measures (TRIMS), trade-related aspects of intellectual property rights (TRIPS), and technical barriers to trade (TBT). The SPS, TRIPS and TBT agreements all exert external demands on policy processes traditionally under the purview of domestic agencies, from regulating health standards to policing counterfeit products. The TRIPS agreement in particular "signaled a new era" for the trading system, given the substantial legal and institutional requirements placed on the state to establish a system of intellectual property rights protection.[3] GATS imposes requirements on the state to disclose information related to services provision and to establish judicial bodies to deal with administrative disputes.[4] The SCM agreement delineates categories of subsidies that are either prohibited (e.g., export subsidies) or actionable (e.g., specific production subsidies), while TRIMS prohibits the use of measures such as local content requirements.[5]

In other words, these WTO agreements do not just impose restrictions on what policies may not be conducted but also actively generate a set of demands for the establishment of regulatory and legal institutions grounded in liberal market economy norms, from health and safety standards to dispute settlement procedures, requiring states to undergo a "fundamental and intrusive restructuring" of domestic institutions.[6] Such restructuring is an inherently political process, given its potential for altering the balance of power across different agencies within the state. These external demands generate pressures to enhance a state's regulatory capabilities, thereby potentially empowering state agencies tasked with standards setting and monitoring, intellectual property protection, and other market regulation functions.[7]

WTO membership also raises the technocratic capacity required for nation-states to participate in international economic diplomacy, from negotiating complicated and highly technical agreements to managing the legal processes of trade disputes. The WTO's dispute settlement mechanism allows member states to bring complaints about the policies of other members before a WTO panel. Certain trade-related agencies within the state's governing apparatus will therefore gain increased responsibilities over bringing such cases as a complainant and, as a defendant, accounting for domestic policies accused of being in violation of WTO rules. These demands again generate the potential for political conflict within the state, as some of the offending policies might have been enacted by other economic ministries or authorities in the government. In the event that a member loses a case, the reputational costs and pressure to comply with panel rulings can generate disagreements between, for example, trade ministries and authorities in charge of industrial policy.

Member governments are further required to reach a high standard of information disclosure regarding their trade and economic policies.[8] The WTO imposes over two hundred notification requirements on members and requires all members to make public their general laws, regulations, and administrative rules. The WTO itself provides comprehensive trade-related data on all members, from tariff and nontariff barriers to in-depth trade policy reviews of country practices. In the event of a dispute, both sides are obliged to present their own evidence at the WTO panel to support their arguments. Information on a WTO member's trade and domestic policies has therefore come to carry high economic benefits, as such data can affect panel rulings. Moreover, meeting the WTO's transparency requirements can potentially alter lines of information flow within the state. This in turn raises the potential for conflict between agencies charged with collecting information and those tasked with furnishing data about their activities for external scrutiny. Such changes are arguably more intrusive and demanding for authoritarian states accustomed to operating in opacity.[9]

Finally, while the GATT regime allowed members certain flexibilities and differential treatment in the application of rules, the WTO requires newly acceding members to sign on to all existing rules in a "single undertaking." This single undertaking was a particularly onerous task for China, as it represented a process of change that was much at odds with the incremental and heterodox nature of reform that had marked its previous three decades of growth. Additionally, Article XII of the WTO Agreement states that a newly acceding member is to join the WTO "on terms to be agreed between it and the WTO," effectively giving current members the power to impose on new members obligations and commitments that go beyond existing WTO rules. In the preaccession stage, the party seeking entry naturally has a weak bargaining position vis-à-vis existing members, allowing the latter to negotiate accession terms far stricter than those imposed on the members themselves. This raises the amount of domestic adjustment that the acceding party has to implement as a price for accession.[10] In no case was this more true than that of China's WTO accession, as the next section discusses.

The Political Economy of China's WTO Entry

China had been an original signatory of the GATT when it was formed in 1947. After the Kuomintang was defeated by the communists in 1949, the Kuomintang fled to Taiwan and subsequently announced China's withdrawal from GATT. In 1986, the Chinese Communist Party (CCP) government began the process of resuming GATT membership.[11] Little progress was made over the next decade not only because of the violent Tiananmen crackdown in 1989 but also due to ambiguity over how to treat China's self-described "socialist commodity economy" within the ambit of GATT rules.[12] Further disagreement lay in whether China would be treated as a developing country, given its size and potential economic import.[13] When the WTO was formed in 1995, China kept up its integration efforts by applying to join the organization as a contracting party. The process of WTO accession, however, would be substantively more difficult, given the enlarged scope of WTO rules described earlier. Negotiations would now have to cover not just liberalization in goods but also services as well as policies governing intellectual property, dispute resolution, transparency, and investment.[14]

Detailed accounts of the events and political dynamics leading up to China's WTO entry can be found in a number of scholarly works.[15] This discussion therefore focuses on why China's accession commitments represented such an external shock to its domestic governing system. The analysis highlights the role that foreign economic powers—in particular the United States— played in shaping the specific commitments in China's protocol of accession. Then, the analysis explains how China's leaders overcame tremendous domestic resistance to WTO entry, why the resulting terms in the protocol went above

and beyond ordinary WTO commitments, and why we might expect bureaucratic resistance to WTO-led liberalization.

US interests played a critical role in shaping the major terms and conditions
for China's WTO accession, because the United States was the most important
gatekeeper to WTO entry. Once China came to a bilateral agreement with the
United States on its accession terms (signed in November 1999), negotiations
with other major trading partners such as the European Union and Japan pro-
ceeded much more smoothly.[16] These US interests derived from a combination of commercial and geopolitical considerations. The commercial interests
were grounded in the economic potential of China's domestic market and exporting capabilities, while the political motivations were driven by a broader
conviction that it was important to knit China closer into the international system and that WTO entry would strengthen the hand of reformers in China.[17]

In terms of economic interests, United States Trade Representative (USTR)
Charlene Barshefsky spent two years of the accession process emphasizing to
the Chinese that China had to accede to the WTO on commercial—not
political—terms.[18] These commercial demands were articulated in a road map
to accession that Barshefsky presented to Chinese trade minister Wu Yi in 1995.
The document set out the key commitments that China would have to undertake to join the WTO, including not just major cuts in tariffs on manufacturing goods but also substantive liberalization in sensitive areas such as agriculture and services (banking, telecommunications, and insurance).[19] Every single
tariff line came under intense scrutiny and discussion, a reflection of the broad
range of US commercial groups interested in gaining access to China's large
domestic market. Therefore, to have the trade deal passed by Congress, the
US negotiating team was under immense pressure to achieve deep and broad
concessions.[20] On the protectionist side, the negotiators also had to incorporate concerns from textiles and steel groups that safeguards be built into the
deal so they would not be hurt by surges in Chinese imports.[21]

As for broader geopolitical considerations, China's WTO entry represented
an opportunity to integrate China with the rules of the international economic
order. As a former US official put it, "In the broadest form, the motivation was
to ensure that China's reemergence would be constructive, not destructive, to
the global system of order."[22] This belief only deepened when the WTO was
formed, since it had stronger enforcement powers via its dispute settlement
mechanism.[23] US president Bill Clinton emphasized this point when drum-
ming up domestic support for this deal, noting in a speech in 2000 that "under
this agreement, some of China's most important decisions for the first time will
be subject to the review of international bodies with rules and binding dispute
settlement."[24] Treasury Secretary Larry Summers reinforced these views during his testimony to the House Committee on Ways and Means in 2000:

> For the first time, China's compliance will be subject to multilateral enforce-
> ment under the WTO dispute settlement mechanism, which will force China to

comply with WTO rulings or be subject to trade sanctions. . . . [WTO member-
ship] will both support the cause of market reform within China and provide
an effective rule-based framework for future Chinese reforms to take place. . . .
By learning to "play by the rules," both internationally and domestically, China
will strengthen the rule of law, which will enable it to become a more reliable
partner and a fairer society.[25]

What were the interests of China's leaders in joining the WTO. and how did
those interests line up with those of China's bureaucracy? One historical reason for China's WTO bid was that its paramount leader, Deng Xiaoping, had
made the decision to reintegrate China into the international system.[26] Broadly,
WTO entry would advance China's efforts to gain recognition as a legitimate
and major power on the international stage.[27] More crucially, a combination
of economic factors made WTO entry a high priority for China's leaders in
the late 1990s. First, the 1998 Asian financial crisis meant that regional growth
prospects were weak and rates of foreign investment in China were declining.
Second, Premier Zhu Rongji's effort to reform the state-owned enterprise
(SOE) sector was running into problems, and he came to see WTO entry as
providing important external pressure to force the Chinese economy (and its
SOEs) to shape up. As Margaret Pearson points out, "the only way to break
the hold of the 'old' economy and its champions was to force change on it via
the stringent requirements imposed by WTO rules."[28]

However, the leadership's goal of WTO entry was heavily opposed by bureaucratic and industrial interests within China. Indeed, resistance to liberalization was so entrenched that a more consultative approach taken in the 1980s
and early 1990s had led to bureaucratic deadlock, requiring a shift to a more
streamlined approach restricted to the top leadership.[29] Industrial ministries
resisted the prospect of having their authority curtailed by international rules
and did not welcome the threat of foreign competition bearing down on the
industries and firms under their charge. Given the sweeping liberalization that
the US negotiators sought as a price for entry, the range of industries resisting the deal was similarly broad: from agriculture and automobiles to telecommunications, chemicals, insurance, and machinery.[30] In late 1997, a number
of central ministries, provincial representatives, and industry managers appealed to the Chinese leadership to postpone WTO accession.[31] The automotive industry also directly petitioned President Jiang Zemin (who had spent
part of his career in the auto sector) not to make such concessions.[32]

This domestic opposition swelled after Premier Zhu failed to secure a deal
for WTO entry with President Clinton on a visit to the United States in
April 1999. The US rejection was a political blow to Zhu, one that was worsened by a backlash in China when the list of Chinese concessions was posted
on the USTR website. The agreement was denounced by critics as a new
"Twenty-One Demands," underscoring the degree to which opponents saw the
deal as a violation of national sovereignty.[33] According to one insider account,

the public posting gave many in the Chinese government their first look at the substantial and sweeping liberalizations that Zhu had offered to the United States, provoking an internal party reaction that was "as strong as a force-six earthquake."[34] One newspaper report at that time noted that negative reactions also spread to local governments in China's interior agricultural regions. The proposed liberalization in agriculture would have opened Chinese markets to imports of US wheat, fruit, and meat, leading local leaders to petition the central government against the deal.[35] Proposed concessions in telecommunications, which included allowing foreign investors 51 percent ownership shares, reportedly led the minister of information industry, Wu Jichuan, to offer his resignation in protest.[36] Officials from the Ministry of Foreign Trade and Economic Cooperation (MOFTEC) came under attack from other ministries (including agriculture) demanding explanations for how and why these concessions were made.[37] Zhu's offer to the United States was widely decried by his ministers for being excessive, leading some officials to accuse him of betraying his country.[38] These reactions laid bare the gap between the leadership's ambitions for China's WTO entry and broader resistance within the bureaucracy.

The negotiations, already hobbled, were almost completely derailed in May 1999 when the US military mistakenly bombed the Chinese embassy in Belgrade as part of North Atlantic Treaty Organization air strikes on Serbia, sparking anti-US protests across many parts of China.[39] Bilateral talks were only restarted after a series of outreach efforts made by Clinton and Summers. Even then, the final negotiations in November 1999 repeatedly led to stalemate in discussions led by MOFTEC minister Shi Guangsheng.[40] This was in part because while MOFTEC was the lead agency heading the Chinese interagency process, it did not have the power to impose its will on the other ministries resistant to making concessions. Hence, individual ministries would hold their ground during the negotiations until Jiang and Zhu stepped in to break the deadlock.[41]

Indeed, given the breadth and depth of domestic opposition, the only way for China to conclude the agreement was for the leadership to intervene directly over the objections of the bureaucracy. The more controversial issues were therefore only resolved when Premier Zhu twice intervened directly in the negotiations.[42] It was a "political war" on the Chinese side, and Zhu faced "enormous political resistance."[43] Although Zhu was not the only reformist within the government, the pro-WTO coalition within China was politically weak. This meant that Zhu's intervention to pull the bilateral deal together was founded on an extremely fragile coalition that could have easily fallen apart.[44] Once the US-China deal was finally struck, however, the road was paved for China to conclude its bilateral talks with other major trading partners such as the European Union and Japan and to join the WTO in December 2001.[45]

Of course, not all groups in the United States supported China's WTO accession, and to say that China was completely unable to protect its interests would be inaccurate. In the United States, opponents to China's WTO accession included human rights activists critical of labor conditions in China and labor unions concerned about threats to US jobs and discriminatory industrial policies. China also pushed back on some of the US demands, particularly in services liberalization. During the final negotiations with the United States, Premier Zhu refused the US request for foreign enterprises to have controlling shares in movie theaters, saying that "everyone knows that ideology is very sensitive; let's leave the controlling share issue to be discussed later!"[46] He also reversed earlier concessions for 51 percent foreign equity shares in telecommunications and insurance on the basis that China's political situation had changed after the Belgrade embassy bombing.[47] Zhu additionally rejected further liberalization in telecommunications, saying that everyone involved within China would not agree to this.[48] Finally, the Chinese also retained state trading in specific products, including chemical fertilizers, tobacco, cotton, oil, grain, and more.

Nevertheless, the US-China bilateral agreement that was finally struck went far beyond what the United States thought the two countries could achieve. In most circumstances, after bringing back a deal to the US Congress, USTR negotiators would be asked by domestic interests to return to the trading partner with additional demands. But in this case not a single industry or senator made that request, revealing the degree to which the US negotiators had succeeded in their task.[49] Reports on the bilateral agreement noted that "rarely has the United States pinned down a trading partner on so many elements of business life."[50] One assessment concluded that the US negotiators had "exceeded expectations."[51] The deal was pathbreaking in its specificity, leading to an accession protocol containing market access commitments that were "unprecedented in WTO history."[52]

China's WTO Commitments

The impressive scope and depth of China's WTO accession protocol was designed partly in recognition of the weak legal institutions in China's opaque and authoritarian political system and also in light of the enforcement power of the WTO: WTO rules have the status of international treaty obligations in China, rendering them more effective than international rules are for other countries such as the United States.[53] In terms of economic liberalization, China's import tariff rates were bound at a low average level of 10 percent, with very little gap between the applied and bound rates, a commitment that distinguished China from other large developing countries such as India (a founding member of GATT) that enjoyed more room for maneuver in their tariff commitments.[54] In addition, China's liberalization of its services sector went far beyond that offered by other countries at the same and even higher levels of income.[55]

China's protocol of accession further included detailed and wide-ranging commitments on how the state would govern key aspects of the Chinese economy, from state-enterprise relations to establishing processes for judicial review and limitations on policy measures ranging from subsidies to foreign investment. China not only agreed to give up many of the exceptions and flexibilities that would normally be accorded to a developing country but also made a number of WTO-plus and WTO-minus commitments going beyond existing WTO rules: the former represents commitments that set a standard of behavior higher than that required by WTO rules, while the latter represents commitments that allow other trading partners to lower the standards they would normally apply in using trade remedies to protect against Chinese exports (these remedies include safeguards and antidumping and antisubsidy measures).[56] In brief, the WTO-plus aspects of the protocol included commitments

- To let the market determine the prices of all goods and services, with a few exceptions such as tobacco, pharmaceuticals, and public utilities;
- To grant trading rights to all entities, Chinese or foreign, within three years— again with a few exceptions in specific categories;
- Not to interfere in the commercial decisions of SOEs;
- Not to add conditions to foreign direct investment approval relating to technology transfer and R&D activities in China (which goes beyond the prohibition of local content requirements contained in TRIMS);
- To expand the existing scope of national treatment to include foreign investors in services, trading rights, and the production and sales conditions of foreign businesses in China;[57]
- To implement transparency commitments such as establishing an official journal publishing all trade-related measures; setting up information points for business inquiries about trade-related measures, to reply to requests for information within thirty days after receiving the request (up to forty-five days for exceptional cases), providing a reasonable comment period for all draft measures, and translating all trade-related measures into one of the official WTO languages;
- To allow enterprises and individuals involved in judicial or administrative tribunals the opportunity for appeal under all circumstances as well as provide written decisions to the appellants and inform them of any rights they might have to further appeal;
- To nullify any subnational measures that are inconsistent with WTO rules and the protocol of accession and to establish a complaint mechanism for notifying the government of nonuniform application of rules; and
- To accede to an additional layer of scrutiny through a transitional review mechanism, which would monitor and assess China's progress in implementing its WTO commitments regularly over a period of ten years.

In addition, the WTO-minus aspects of China's accession protocol included commitments

- That WTO members can treat China as a "nonmarket economy" in calculating domestic prices in antidumping cases for a period of fifteen years;[58]

- That WTO members can use a similar nonmarket economy methodology in determining the existence of subsidies under the SCM agreement, with no expiration date;
- That for a period of twelve years after accession,
  - WTO members can trigger safeguards against Chinese imports in cases where these imports cause "market disruption" (the existing safeguards mechanism sets a higher standard, requiring imports to be "a significant cause of material injury"),
  - China can only retaliate against these safeguard measures after a period of two years,[59]
  - These safeguard measures can be applied for any period of time (exceeding the eight-year limit under WTO rules), and
  - Members worried about a trade diversion effect caused by the imposition of a safeguard measure by another member against China can launch an investigation into triggering their own safeguard immediately;
- That until the end of 2008, members can trigger special safeguards on Chinese textile imports as long as there has been "market disruption threatening to impede the orderly development of trade in these products" and that China cannot retaliate against these safeguards; and
- That unlike the treatment accorded to developing countries,[60]
  - China's subsidies to SOEs are to be treated as "specific" subsidies and hence would be actionable under the SCM agreement,
  - China's debt forgiveness programs would be actionable under the SCM agreement, and
  - De minimis support in agriculture would be capped at 8.5 percent of the total value of production of a particular product, lower than the average 10 percent held by developing countries.

Together, these commitments added up to no less than a herculean undertaking to transform the functioning of the state in China's domestic governance system. Why did China agree to such a high price of entry? First, as it was the Chinese who were seeking access to the organization, much of the leverage during negotiations naturally lay with the United States.[61] Second, China's leadership sought to use WTO entry as a credible commitment to push domestic reform forward. These political dynamics meant that the terms of entry were agreed upon over the strenuous objections of a bureaucracy that then was charged with implementation. The task of implementation would clearly involve fundamental shifts in state-market relations and the functions of government and threaten to undermine the interests of existing bureaucratic actors. As such, it is not surprising that the potential impact of this external shock on the party-state's governance was a topic of intense discussion within the CCP. In 2001 and 2002, several articles on the impact of WTO entry on Chinese governance and sovereignty appeared in journals published by the Central Party School such as the *Chinese Cadres Tribune* and the *Central Party School Journal*. In one article featuring an interview with Chen Qingtai, the deputy director of the State Council's Development Research Center, Chen pointed out that "China's entry into the WTO raises many challenges, amongst which the most direct and salient is the challenge it brings to the government's administrative structure and behavior."[62] In another article Li Junru, the deputy head of the Central Party School, explained this challenge in greater detail. He noted that WTO entry

> is an important strategic move for our country's economic development but also brings along with it a problem, which is how the Party, in its decision and policymaking process, can defend China's national interest and socialist principles while fulfilling international norms and WTO rules. The biggest "shock" of the WTO to China is aimed at our government, and we are a government governed by a single party, so in reality the ones who will feel the "shock" will be the Party's administrative and leadership methods.[63]

How would China's party-state respond to these terms of liberalization? Given the substantive and intrusive demands imposed by China's accession protocol, what are the ways in which we can assess the politics behind Chinese policy responses to WTO entry?

Clarifying the Relationship between International Rules and Bureaucratic Politics

Surprisingly, the question of how WTO entry affects politics within the state has been relatively neglected in the literature. Existing scholarship on international trade institutions has focused on issues such as institution formation, legal adjudication, the economic benefits of membership, and the politics of accession.[64] Relatively less work examines the political process through which the greater intrusiveness of international trade rules might alter domestic economic policymaking.[65] One 2014 literature review titled "Governance from the Outside In" even noted that "we know of almost no research on the impact of PTAs [preferential trade agreements] on the structure of domestic institutions."[66]

Indeed, the expanding scope of international economic rules is often an omitted variable in international political economy scholarship, which has been focused instead on "open economy politics."[67] As a result, studies on the political economy of trade have tended to be framed in dichotomous terms, with a focus on protection as the main alternative to liberalization, and demands for protection premised on economic interest groups organized along class lines or sectoral cross-class coalitions.[68] In these frameworks, the state is treated as primarily balancing between the desire to raise overall social welfare through liberalization on the one hand and responding to demands for protectionism by economic interest groups on the other.[69]

Studies that do focus on the effects of international rules on state behavior have largely been concerned with the issue of compliance, with extensive and important work examining the various mechanisms that might influence the probability of a state complying with or defying its international obligations.[70] In this vein, several studies have examined the question of whether China is complying with WTO rules. Some have found China to have a strong record of compliance and point to the role of reputation costs in driving this compliance.[71] Others argue that this compliance behavior is in form rather than substance and belies the active introduction of new WTO-inconsistent measures.[72]

In research specific to the WTO, analysis has gone beyond the compliance question but nonetheless focuses on a "yes or no" debate about whether the more binding rules of the WTO have constrained the policy space of developing countries. Those arguing for the WTO's constraining effects point to how its rules restrict the use of policies that economically successful countries have long relied on for growth.[73] Scholars opposing this view argue that WTO restrictions still leave plenty of alternative measures—such as trade preferences, phase-in periods for certain commitments, and safeguards—for developing countries to achieve their goals. Moreover, they argue, a wide range of domestic policy domains, such as R&D and nonspecific subsidies and foreign investment, remain relatively untouched by WTO rules.[74]

This focus in the literature on dichotomous outcomes—of protection versus liberalization, compliance versus defiance, and constraint versus nonconstraint—obscures the fuller range of potential outcomes that might result from entry into an international organization. States rarely face a simple choice of either abiding by or resisting the demands of global rules and domestic interest groups. Instead, they seek to further their interests by engaging both external and internal environments simultaneously, choosing responses from a "great variety of instruments and adaptation strategies" that are available to them.[75] This means that states might face a choice across multiple policy responses that all fit into the broad category of noncompliance. However, choosing among these policies involves deciding between measures that bear distinct political and economic implications.

Additionally, such dichotomous framings tend to discount the likelihood that more than a single policy response might issue from the state, possibly driven by disparate actors within the bureaucracy's diverse arms and functions. By ignoring the bureaucratic separation of responsibilities for formulating economic policy, these frameworks obscure how global trade rules differentially affect the interests of distinct governmental actors. As this chapter has emphasized, WTO rules have the potential to threaten the influence and autonomy of some bureaucratic agencies by reducing their discretion over the ways in which they can govern specific aspects of the economy, from setting quotas to the provision of subsidies. At the same time, other bureaucratic actors might potentially see opportunities for expanding their influence with WTO entry, considering the degree to which WTO rules increase the demand for a technocracy well versed in international trade law. In other words, external rules may substantively alter politics *within* the state, affecting the interests of specific actors differentially.

How might WTO entry have affected bureaucratic interests in China specifically? One conceptual difficulty in bringing the international context into the politics of China is the size and complexity of its governing structure. Administrative governance in China is decentralized into four major subnational layers: the province (alongside special administrative regions such as Hong Kong and Macau), prefectures and prefecture-level cities, counties and county-equivalent units, and townships.[76] From the late 1970s onward, as China moved away from centralized planning, much authority over economic governance was devolved to these subnational units. This decentralization generates challenges, because tangled incentives and conflicting, multilayered accountability relations generate substantial scope for different bureaucratic actors to deviate from the instructions of higher-level authorities.[77] Moreover, the diversity of socioeconomic conditions across different regions in China generates immense differences in governance challenges and priorities at the subnational level.[78] Consequently, national-level policies are often thwarted or distorted by implementation and enforcement problems on the ground.

Even within the central government, powerful agencies jostle with each other for influence and control over the economic agenda. Although the premier formally oversees the economic portfolio, multiple ministerial-level agencies have jurisdiction over different (but overlapping) aspects of the economy. For example, while the Ministry of Industry and Information Technology is in charge of high-tech industrial policy, the National Development and Reform Commission, the Ministry of Commerce, the Ministry of Science and Technology, the Chinese Academy of Sciences, and the Chinese Academy of Engineering are also likely to have various degrees of policy authority and influence over the same issue area. As a result, multiple actors in different parts of the bureaucracy all have a say in—and compete with each other over—the direction of economic policymaking, while authoritarian rule endows them with disproportionate influence over how an industry develops.

One consequence of this "fragmented authoritarianism," as Tony Saich points out, is that analyses of the Chinese state at times resemble the parable of the blind man and the elephant, with different scholars homing in on different parts of the sprawling and variegated party-state.[79] Each scholar emerges with their own pronouncement of the political dynamics at play depending on which element of the colossal governing structure is being studied. Studies assessing the impact of WTO entry on Chinese industrial policy very much reflect this challenge of systematically capturing complex dynamics within a large political entity. China's WTO accession in 2001 provoked an early set of studies predicting the effects of WTO membership on the Chinese political economy. Some examined the ability of the Chinese state to fulfill its WTO commitments, while others discussed the potential distributional impact of trade liberalization in terms of sectoral unemployment and regional income inequalities.[80] Still others analyzed the impact of trade liberalization on key strategic industries such as telecommunications and automobiles.[81] Another set of later literature investigated China's behavior within the WTO itself, centered around the question of whether China has become a status quo player on the international stage.[82]

Relatively fewer studies have focused on the impact of WTO entry on China's domestic institutions.[83] For the most part, the global economy is acknowledged as a passive background variable by reference to a general opening of economic borders. This omission is an increasingly consequential one, as China's membership in various international institutions expands and as the scope of these institutions gradually reaches behind the border to directly impact China's domestic governance structure. Studies that do take a more institutional focus have homed in on partial—rather than holistic—aspects of economic governance, leading to contradictory findings. For example, some studies have noted the creation of new legal and regulatory institutions, the liberalization of large portions of the economy, and a significant reduction in the number of SOEs in the WTO era.[84] Yet other studies have highlighted the rise of policies promoting strategic national pillar sectors and a strengthened state activism.[85] One set of scholars focused their lenses on regulatory reforms, and another zoomed in on state intervention in strategic industries. Each captured a dynamic that was part of a larger complex system.

While some studies do adopt a wider lens to capture broader variation within the governance structure, they tend to argue or assume that the policy patterns are the result of a unified, coherent state design. For example, Roselyn Hsueh explains variation in the degree to which different industries have been liberalized in the WTO era, positing that the state adopts a dual-track approach of strengthening its control over strategic sectors while allowing for liberalization in less important industries.[86] Similarly, Margaret Pearson argues that Chinese economic governance comprises a three-tiered structure with different governing policies for industries in each tier.[87] These frameworks, while compelling, tend to overlook the intense contestation between different parts of the state over the appropriate measures with which to govern the economy. How do competing interests within the party-state come to an agreement over the direction of economic policy? If there is broad-based agreement within the party-state over how to govern the economy, what explains changes in the policy trajectories of similarly strategic industries over time?

This study sees the tensions and heterogeneity in China's economic policies not as part of an intentional or coherent state design but rather as the product of politics within the state. As the previous chapter discussed, China's fragmented authoritarian system regularly generates policy conflict between different parts of the state, not just vertically between central and subnational governments but also horizontally across different agencies within a given administrative level.[88] These diverse internal state interests mean that the demands on the state imposed by WTO rules were experienced differently depending on where one is located within the state structure. There is thus little reason to expect a convergence toward liberalization, or a unified state response, to WTO entry. Moreover, the resulting policy responses may not map onto a simple dichotomy of liberalization versus protection. This is because China's process of economic reform has long been marked by deeply entrenched internal disagreements over how best to govern the economy, as the next section discusses. Even as Chinese leaders negotiated their wide-ranging commitments to the WTO, the previous decades of reform had involved the embrace of a wide set of governance approaches such that tools available to various governmental actors within their policy repertoire ranged from liberal market economy tools to Japanese-style industrial policy and Maoist-era command economy directives.

Chinese Industrial Policy and Reform in the Pre-WTO Era

China's evolution away from the planned economy involved the introduction of new governance approaches that supplemented but never fully replaced preexisting modes of policymaking. This uneven process of liberalization reflects long-standing internal debates over the appropriate goals and instruments of reform.[89] During the Mao Zedong era (1949–1976), China's economy operated through central planning and campaigns, with economic development skewed toward heavy industry and driven by military strategic considerations. The economy was highly insulated, with trading rights restricted to twelve state-owned trading enterprises and a foreign exchange system that was fixed and nonconvertible.[90] The combined damage of the Great Leap Forward and the Cultural Revolution substantially delegitimized Maoist approaches to economic governance and made the revival of economic growth a critical priority for the CCP when Deng Xiaoping took over as de facto leader. *How* the economy should be governed, however, was a question subject to broad disagreements. As part of the internal debate, Chinese reformers looked outward and studied practices in East European countries such as Hungary and the developmental experiences of Asian countries such as Japan, Korea, and Singapore. While the Japanese and Korean models came to be the highly influential, Western market economy models were also scrutinized.[91] Drawing from all of these external cases and lacking any clear strategy, Deng-era reforms were carried out on a tentative experimental basis, incrementally exploiting opportunities for change outside of rather than within the planned economy.[92]

Agricultural reforms took place through the Household Responsibility System that broke away from the commune and allowed production to be contracted to the household. In manufacturing, much production was decentralized to township and village enterprises.[93] Dual-track pricing allowed for the retention of planned targets, while prices above quotas were liberalized. Experiments with trade liberalization were launched by designating four cities as special economic zones. These reforms ignited economic growth through an expansion of the nonstate sector but left the core of the planned economy largely in place.[94]

As these experiments paid off, efforts to consolidate the direction of reform culminated in the adoption of the "Decision on Economic System Reform" at the Third Plenum of the Twelfth Central Committee in 1984. Rather than jettisoning the command economy, the decision stated that the goal of reform was to establish a "planning system that consciously applies the law of value, to develop a socialist commodity economy."[95] The term "socialist commodity economy" allowed China's leaders to forge a compromise between a planned economy and one that operated on a market economy with commodity exchange, thereby creating space for Japanese and Korean models of development.[96] The 1986 Seventh Five-Year Plan made the first mention of the term "industrial policy," marking the introduction of a new policy approach within the ambit of the planned economy.[97] The following year, the State Council's Development Research Center drafted the "Preliminary Study of China's Industrial Policies," noting that "the experience of Japan and South Korea, as well as other countries and regions, in using industrial policy to achieve an economic system that combines 'competition' and 'intervention,' is deserving of our serious attention."[98] At the same time, the Industrial Policy Department was established in the State Planning Commission, reflecting growing space within the planning bureaucracy for the adoption of East Asian–style development approaches.[99]

The country's first "Decision on Current Industrial Policy Priorities" was issued in April 1989.[100] Rather than a full embrace of Japanese-style industrial policymaking, the 1989 decision contained a mix of contrasting approaches to economic governance, emphasizing the need to "organically combine planning with market."[101] For example, while the decision advocated the use of credit, taxation, and pricing policies, it also retained a heavy emphasis on state control over foreign exchange and foreign capital. Reflecting the patchwork of ideas driving this national policy, the document also combined market economy terms such as "macroregulation," "market demand," and "economic efficiency" with traditional Marxist concepts such as "fields of social reproduction" and production targets. The direct effects of this policy are difficult to assess, as the Tiananmen crackdown two months later would throw the country into a political crisis and introduce a period of stasis and even retrenchment in terms of economic liberalization. Internal party debates over the role and scope of reform—and indeed the relationship between the plan and the market—intensified.[102] There would be no bold policy changes until after Deng Xiaoping undertook his famous Southern Tour in 1992 to urge localities to once again embrace reform and opening.

Deng's Southern Tour cooled debates over reform and paved the way for a consolidation of—but not consensus over—policy direction. In 1992, the CCP's Fourteenth National Congress affirmed that the goal of China's reform was to establish a "socialist market economy." The supporting "Decision on Several Questions relating to the Establishment of the Socialist Market Economy" was issued at the Third Plenum of the Fourteenth Central Committee the following year, setting forth a new round of reforms that brought heightened emphasis on market-oriented practices.[103] Rather than a full embrace of market economy principles, however, the policy consolidation involved a simultaneous adoption of market and statist economic principles. On the one hand, the 1993 decision explicitly sanctioned the building of new institutions to support market economy functions, from centralizing the central banking and fiscal systems to unifying the exchange rate and reforming the SOE sector, signifying a shift toward establishing common rules for governing the Chinese economy.[104] On the other hand, while emphasizing SOE reforms such as enterprise modernization and a separation between government and enterprise, the document simultaneously affirmed state ownership as the "principal component" of the economy and noted that large SOEs held the position as "pillars" of the national economy. Prominent economist Wu Jinglian, in his own assessment of the mixed messages emanating from the party, noted that "although the goal of reform was clearly to establish a 'socialist market economy,' there were different interpretations of what a 'socialist market economy' actually was" and that "this lack of precision allowed different understandings of the objective of China's reforms to persist."[105]

The competing interpretations of China's reform goal can be seen by the divergent conclusions drawn by different ministry officials regarding the developmental experiences of Japan and Korea. For example, Huan Guoyu from the State Economic System Reform Commission (a liberal-leaning outfit formerly headed by General Secretary Zhao Ziyang), studying Japan and Korea's growth experiences, concluded in a 1996 article that while government intervention was needed in underdeveloped economies, this intervention should seek to compensate for and support the market, not destroy the market. Effective governance therefore required a transformation of government functions as well as enterprise reforms to encourage profit maximization under hard budget constraints and with independent operations.[106] In contrast, Chen Bingcai from the conservative State Planning Commission, in an article that same year analyzing South Korea's conglomerates, noted that the key benefits of supporting conglomerates were that large enterprises could diversify risks and undertake long-term plans requiring large economies of scale and, crucially, that these enterprises could be relied on to effectively implement government policy.[107] Hence, while one official saw in the Japan/Korea experience the necessity of separating enterprise-government relations and establishing enterprise autonomy, another saw the benefits of deploying enterprises as a tool of government policy.

These competing approaches to economic governance found their way into national policies. In 1994, the State Council issued a revised framework for industrial development in its "Outline of State Industrial Policies for the 1990s."[108] Like its 1989 precursor, this outline combined a range of governance concepts, acknowledging the "fundamental role of the market in resource allocation" on the one hand while emphasizing the need for "the state's macroeconomic regulation and *control*" on the other and further identifying "pillar industries" such as petrochemicals, automobiles, and electronics that would enjoy special state support.[109] The outline further betrayed a continued reliance on planned economy methods by including extensive production targets for different industries in its appendix. For example, the target for sedans states, "For new, expansion and reconstruction/expansion projects as well as technology-upgrading projects for automobiles with engine capacity below 1,600cc: at least 150,000 units."

These mixed messages persisted in economic policies issued in the late 1990s and early 2000s, where important changes were made that strengthened the role of SOEs yet also boosted the status of private entrepreneurs. In 1999 as negotiations for China to enter the WTO entered their final phase, the constitution was amended to formally recognize private ownership and the rule of law.[110] Yet that same year, the Fourth Plenum of the Fifteenth Central Committee issued a "Decision on Major Issues concerning the Reform and Development of State-Owned Enterprises" that reaffirmed state ownership as the "principal component" of the economy and SOEs as its "pillar," signaling policy continuity from the 1993 Third Plenum Decision. The 1999 decision further clarified that SOEs would retain control in a few sectors—national security industries and natural monopolies, industries providing public goods and services, and "pillar" industries—and hold the position of "backbone enterprises" of new high-tech industries.[111] While affirming the importance of state ownership in the economy, Jiang Zemin made an important move to alter the political status of private entrepreneurs through his new concept of the "Three Represents." This ideological change opened up space for private entrepreneurs to join the CCP and was formally incorporated into the party constitution at the Sixteenth Party Congress in 2002.[112]

This short discussion of the various milestones in China's reform process makes clear that contestation between different modes of economic governance has been a constant theme in China's economic story. While the leadership has made important political decisions at different points in time to consolidate the direction of reform, oftentimes the resulting policy documents or party declarations have reflected, rather than resolved, deeply entrenched tensions over how the state should govern the market. Ambiguity over the meaning of broad concepts such as "socialist market economy" allowed for these disagreements to persist but politically also served to paper over differences without derailing the overarching objective of growth. China's economic progress over the past several decades should therefore be seen not so much as the product of liberal market reform or an overarching strategic design but rather the product (intentional and unintentional) of different actors within the party-state drawing from a mixed bag of approaches and accompanying policy instruments ranging from heavy-handed administrative directives to arms-length regulation.

Despite these haphazard and sometimes halting trajectories of reform, however, the Chinese economy underwent substantial transformations in the post-Mao years running up to WTO entry but stopped far short of full market liberalization. From 1978 to 1993, the share of the state sector in industrial output fell from 78 percent to 43 percent, while over 90 percent of prices were liberalized.[113] Market forces expanded in the trading system as well, with the export-to-GDP ratio rising from 5 percent in 1978 to 20 percent in 2000.[114] Following the success of the special economic zones, the government allowed all cities along the eastern coast to engage in export processing trade on terms similar to those enjoyed by the original zones. The number of state-trading enterprises was expanded widely beyond the original twelve, and direct trading rights were also awarded to numerous enterprises. Nevertheless, many parts of the economy remained only partially liberalized even by the late 1990s. While exports had become a large component of national growth, the trade regime remained dualistic, with coastal cities engaged in export promotion while the rest of the country operated under a protected trade regime with high tariff rates and multiple nontariff barriers.[115]

This chapter began by discussing how the WTO established a new set of global rules on trade that substantially encroached into policy areas previously considered to belong to the domestic domain. Not only did China have to sign on to these rules in a single undertaking, but its accession protocol also layered on commitments for policy and institutional reforms that went far beyond ordinary WTO obligations. These deeply binding terms were driven on the one hand by a US government seeking to knit China closer into the liberal international economic order and on the other hand by a Chinese leadership seeking to use external rules to spur domestic reform. The dominant expectation was that WTO entry would serve as a credible commitment to put China irrevocably on a path of market reform. However, whether this convergence to liberalization would take place was far from given. As this chapter has emphasized, disagreement over economic governance is deeply embedded in China's fragmented one-party system. These lines of conflict run not just between agencies in the central government but also between central and subnational authorities. While China's WTO commitments had the potential of empowering some parts of the Chinese bureaucracy, many state agencies resisted the prospect of WTO entry. How would different actors within the party-state respond to WTO entry? Given the heterogeneous and competing approaches to economic governance embedded within the state, what might the range of potential policy responses to WTO entry be? How would WTO rules affect the relative balance of power between agencies and the setting of industrial policy? The next chapter presents my theoretical framework for answering these questions.

Chapter 2

A Theory of State Strategies under Global Rules

This book argues that policy responses to World Trade Organization (WTO) entry are mediated by state structure and politics between substate actors, leading to divergent policy trajectories even under conditions of authoritarianism. To introduce this theory, I first disaggregate China's sprawling party-state into the relevant actors influencing economic policy and explicate the fragmented structure of policymaking within which these actors operate. The theory then brings in the dual pressures for adjustment introduced by WTO entry, which involve not just economic liberalization but also new governance demands. After discussing the economic and institutional channels through which WTO entry alters the interests of different political-bureaucratic actors, I then introduce the range of strategic options available to these actors in responding to WTO entry. The two main variables driving an actor's strategy choice, I propose, are the unevenly distributed likelihood of being sanctioned for deviating from WTO rules and the varying prospects for political advancement. The specific way in which these costs of sanction and benefits of advancement manifest themselves shifts, depending on which dimension of China's governance structure is being considered. The chapter then systematizes these dynamics with a set of hypotheses predicting how an actor's choice of strategic response to WTO entry varies across administrative levels, within the central government, and across industries.

Policymaking in China's (Dis)unitary State

This analysis starts from the well-established premise that although China is governed by a single party, authority over economic policy is highly dispersed within the vast bureaucracy. This means that a multiplicity of substate actors has a stake in the consequences of China's accession to the WTO and that

WTO-led liberalization is likely to affect the interests of these actors differentially. Consequently, we need to consider the possibility that more than one response might emerge from the state in reaction to an event such as WTO entry, driven by different actors within China's party-state.

I focus in particular on the divergent incentives faced by two sets of political-bureaucratic actors. The first are the authorities found across different administrative levels in China's hierarchical bureaucracy, that is, the central and subnational governments. These units are conceptually important because China's decentralized governance structure delegates policy responsibilities over many socioeconomic matters downward to provincial and local levels, leaving subnational governments with substantial autonomy and discretion. The second set of actors are the political leadership of the Chinese Communist Party (CCP) and the major economic agencies within the central state. These agencies play an influential role in setting industrial policies and compete intensely with each other for influence.

One consequence of China's decentralized governance structure is that accountability relations within the state tend to be weak, diffuse, and conflicted by overlapping lines of authority. Given China's geographic size and diversity of socioeconomic conditions across subnational regions, the priorities of any given subnational government are likely to diverge from those of the central authority. At the same time, the central government's low capacity for monitoring and enforcement means that subnational authorities are often able to enact policies that deviate from national goals.[1]

This fragmented accountability within China's party-state affects not just policymaking between the central and subnational governments but also at the top echelons of power—specifically between the party leadership and the central bureaucracy. As Susan Shirk points out, there is a "reciprocal accountability" between party leaders and the rest of the bureaucracy because while the party leadership appoints officials at lower levels, it is those same officials—specifically those in the Central Committee—who are the ones authorized to select or ratify appointments to the leadership. As such, instead of a strict top-down hierarchy that one might expect to exist in an authoritarian regime, "government officials are both the agents and constituents of the party leaders. . . . Officials hold their positions at the pleasure of the party leadership, but party leaders hold their positions at the pleasure of the officials in the selectorate."[2]

The claim that China's central government has weak accountability and enforcement power might seem counterintuitive. After all, China's party-state, having achieved a remarkable growth record that lifted millions out of poverty and withstood the major political upheavals and economic disasters of the Mao Zedong era as well as the 1989 Tiananmen crisis, might reasonably be seen as having a more robust state capacity compared to other developing countries. However, this achievement has to be viewed in light of the monitoring challenges generated by the sheer size of the CCP (ninety million members in 2019);

the enforcement problems resulting from the delegation of economic authority to more than thirty provinces, over three hundred cities, and thousands of counties; and the conflicted authority relations between lower-level units and their higher-level supervisors.[3]

The central government's enforcement power therefore has to be assessed in the context of the limitations of its ability to enforce all issues at all times and the massive scope of its governance challenges. As Will Norris points out, the degree to which a state is able to act in a unified manner is a variable rather than a constant.[4] In the case of China, policy coherence can be undermined by a number of factors—central-subnational conflict, bureaucratic decentralization, and factional competition within the party—all of which, as the following chapters will show, affect the ability of the central state to coherently implement and enforce national objectives. Although the central government instituted a number of policy measures to strengthen its ability to implement China's WTO commitments, these measures have been met with resistance and foot-dragging at subnational levels.[5] These structural issues mean that central enforcement in China takes place largely through managed campaigns, where concentrated punishments are focused on particular issues on a temporary basis either to deal with a crisis (e.g., the SARS epidemic of 2002–2003) or as signals of the center's political priorities.[6] That is, the center is "selectively effective," able to exert enforcement power for a narrow set of issues for a limited period of time.[7] Single-party rule in China therefore does not translate into smooth top-down policy implementation, as one might expect. Instead, actors within the state can deviate from policy goals set at higher levels (e.g., the party leadership, the central government, and even internationally at the WTO) to pursue their own objectives.

The weaknesses of the CCP's internal accountability relations might arguably not be a major issue if political interests within the party-state were aligned. However, a key feature of China's authoritarian governance structure is that members within the party-state do not necessarily share the same goals when it comes to economic governance, nor do they possess the same means for reaching those goals that are shared. Instead, authoritarianism means that political competition is almost entirely internalized within the vast party-state apparatus. To be selected into one of the top positions in the party leadership (e.g., the Politburo Standing Committee), cadres have to start at the bottom ranks and work their way up. As a result, officials in leadership positions compete vigorously with each other for advancement within the administrative hierarchy. The cadre evaluation system, on which promotion decisions are made, places disproportionate weight on economic growth figures achieved by individual leaders.[8] Government leaders in various townships, counties, and cities are then highly incentivized to channel their policies toward maximizing economic growth in their jurisdiction.[9]

This fierce bureaucratic competition manifests itself not just across different administrative strata but also within the central government. Bureaucratic actors in the Chinese system cannot be said to have (or even seek) autonomy in the same sense as in developed democracies.[10] Nonetheless, various central agencies are locked in protracted contests for policy influence.[11] As chapter 1 noted, bureaucratic responsibilities over economic policy are split across numerous agencies. Although the functional scope of each agency appears clearly delineated on paper, in practice policy responsibilities overlap, leading to interagency competition for influence and control over the economic agenda.

This internalization of political competition within China's single-party system means that various substate actors—governments at different administrative levels and central agencies—are constantly searching for avenues to enhance their rank and influence in the bureaucracy and are willing to deploy a wide variety of measures in pursuit of political advancement.[12] The fragmentation of accountability relations within the state further means that actors often have the scope to adopt policies that deviate from those desired by leaders at higher ranks.

When it comes to assessing the impact of WTO entry, we therefore need to consider how the demands introduced by WTO membership are filtered through these entrenched features of policymaking in China. The next section discusses the new economic and bureaucratic conditions introduced by WTO entry before moving on to consider how these conditions affect the political calculus of various substate actors.

New Conditions of Policymaking under WTO Membership

As noted in chapter 1, China's entry into the WTO in 2001 involved wide-ranging commitments that had the potential to dramatically transform the economy and state-market relations. These commitments involved a sharp reduction of tariff rates in manufacturing and agriculture as well as substantial market liberalization in services. The terms of accession included not just a single undertaking to implement existing WTO rules but also a set of WTO-plus and WTO-minus obligations involving additional conditions for entry. Moreover, the package of concessions that China agreed to was negotiated over the vigorous objections of a bureaucracy that would now be in charge of implementation. What would the main concerns of these bureaucratic actors be as they approached the task of WTO implementation? Broadly, I divide the WTO-related concerns of the bureaucracy into two categories: economic effects arising from trade liberalization and bureaucratic effects stemming from new external commitments on state behavior.[13]

As has been well established in the literature, trade liberalization creates winners and losers, generating export opportunities for some and import competition for others and potentially raising overall economic welfare but also introducing threats to industry-specific firm activity and employment. The standard literature on the political economy of this liberalization focuses on economic interest groups, delineated in terms of export promoters versus import competitors, or capital versus labor and land. When it comes to the state's interests in the political economy of trade, the dominant approach is to model the government as run by politicians seeking to maximizing their interests, balancing between contributions from protectionist groups who stand to lose from liberalization and the overall welfare of its voters (which would generally be enhanced by liberalization).[14] However, such models of state behavior are grounded in democratic frameworks and may not be easily extended to authoritarian states. In the absence of electoral pressures, what determines an authoritarian state's interests in the economic effects of trade liberalization? Thus far, these questions have remained relatively neglected in the literature.[15]

I propose that in the context of China, the party-state's interests in the economic repercussions of trade liberalization can be broadly conceived as comprising three aspects. First, in the absence of an electoral mandate, some authoritarian regimes might seek popular support and legitimacy through economic performance.[16] Therefore, to the extent that trade liberalization is linked to enhanced economic growth, there are reasons for the Chinese state as a whole to be protrade.[17] Second, as discussed earlier, economic performance plays an influential role in affecting the advancement of Communist Party leaders at all levels of the bureaucracy, making growth rates a key component of political competition within China's authoritarian system. The extent to which these cadres would be pro- or antiliberalization therefore depends on whether their specific jurisdictions stand to benefit economically from trade liberalization. Third, the share of state ownership in China's economy was still relatively high in the pre-WTO period, despite efforts to privatize, merge, and shut down many unprofitable state-owned enterprises (SOEs) from the mid-1990s onward.[18] Therefore, many state agencies at multiple levels of government remained direct owners of substantial sections of the economy and are more likely to be opposed to liberalization for industries threatened by import competition.

In addition to these economic considerations, WTO entry introduced a set of external rules and commitments to the policymaking process in China specifying the scope of government conduct for issues ranging from tariffs and quotas to subsidies, intellectual property, the rule of law, information disclosure, and more, accompanied by the threat of punishment via the dispute settlement mechanism. China's protocol of accession added yet more commitments over issues such as the conduct of relations between the government and SOEs, technology transfer, and judicial review. The introduction of these commitments therefore generated new pressures on government actors to modify their existing governance approaches and establish new institutions.[19]

Such pressures are arguably more strongly resisted by authoritarian states compared to their democratic counterparts, given that authoritarian bureaucracies, used to operating in relative opacity and without democratic checks on their behavior, would be less willing to cede their discretion over policymaking to an external body.

However, not all actors within a government would be opposed to the new rules introduced by WTO membership. China's commitments included demands on the one hand for building new market-enhancing institutions and formulating arms-length state-market regulations and requirements on the other for the elimination or restriction of various policy measures. Consequently, WTO rules have the potential to both undermine and advance bureaucratic interests, depending on the actor involved. Rules on behavior that is to be eliminated or restricted threatens the influence and autonomy of bureaucratic agencies by curbing their discretion over how to govern specific aspects of the economy. Yet other rules, focused on the creation of new institutions or regulatory structures, potentially increase the demand for regulatory and legal agencies and for a technocracy well versed in a range of issues from standards setting to intellectual property and international trade law. Therefore, bureaucratic actors oriented toward regulatory, technocratic, and legal affairs are likely to perceive WTO entry as generating opportunities for them to expand their influence over policymaking. In other words, the institutional demands of WTO membership generate another set of winners and losers who exist not in the economy but rather within the state apparatus.

Given these cleavages and the economic and bureaucratic channels through which deep integration-style trade agreements directly affect the interests of different actors within the state, how can we conceptualize the range of potential responses that these actors might adopt to WTO entry?

Competing State Strategies

I argue that Chinese bureaucratic actors can choose from a range of different state strategies in responding to WTO rules, where a state strategy is defined as a set of policy instruments, rooted in a particular mode of state-market relations, that are adopted to further a state actor's economic objectives. China's predominant economic goal since the start of post-Mao reforms in 1978 has been economic growth, an objective that has reached into all parts of the party-state and is closely intertwined with incentives for political advancement. At the same time, the state strategies that government actors can choose from have expanded over the course of reform and market opening. As chapter 1 emphasized, China's economic success was not propelled by a single coherent strategy. Rather, it is the product of a combination of gradual adjustments and bold trials drawn from competing approaches to economic governance. One implication of gradualism is that old rules (and accompanying policy measures) persisted in areas where change would be administratively and politically difficult. Distortions in the economy were allowed to remain, but resources were channeled into pockets of less regulated activity.[20] This heterodox approach to reform therefore means that new policies have had to work around, rather than replace, established modes of governance.

This policy gradualism has produced not just uneven trajectories of change but also a diversity of governance approaches within the state. New reforms have tended to be driven and accompanied by the creation of new agencies or elite Leading Small Groups, working around rather than replacing existing agencies that may hold different approaches to reform.[21] These ministries and departments do not fade away and indeed can rise to recover some of their policy influence at opportunistic moments. This layered pattern of institutional change has generated policy heterogeneity and potential conflicts within the state over which types of measures to deploy in governing the economy.[22] I propose that we can conceptualize the diverse range of policy measures as belonging to three types of state strategies.[23] While these strategies share a common objective of fostering economic growth, each is premised on different state-market relations and involves a distinct set of modalities, as explained below:

The *directive strategy* refers to a set of policies based on the traditional tools of the planned economy, such as those found during the Mao period. This strategy, drawing on methods used by the Soviet Union, relies on state command and control over resource allocation and centralized production planning and eschews the use of market mechanisms. Measures issued using a directive strategy will therefore be market-substituting in nature and can still be found in the post-Mao economic system. These include administrative measures such as production targets, directives to shut down factories below a certain scale, or directed mergers to form larger production units. In a free market, decisions about issues such as firm size or closure would not be directed by the state but would instead be made by private firms on the basis of price signals and broader market conditions.

The *developmental strategy* is most closely associated with an autonomous and technocratic bureaucracy charged with mobilizing the resources necessary for industrialization, as seen with the cases of Japan and Korea. I distinguish the classic notion of a developmental state from this developmental strategy in that a strategy is composed of a set of policy instruments and does not necessarily add up to the coherent alignment between economic agencies, firms, and society described in Chalmers Johnson's work on the developmental state.[24] In the Chinese context, the developmental strategy is employed by agencies using state intervention to promote the growth of specific industries but in ways more amenable to open economy competition. Unlike the directive strategy, which dictates economic outcomes through explicit targets, quotas, or direct intervention, developmental policies are premised on market-shaping measures. On the one hand, these measures provide incentives for firms to increase participation in a specific sector. On the other hand, these measures also impose conditions so that the state can access resources such as foreign technology. Examples include equity limits on foreign direct investment (FDI), tax incentives, or subsidies for certain types of investments. As key decisions are still made by firms, these policies are not market-substituting. However, developmental measures are market-shaping in that state incentives nevertheless influence firm decisions.

The *regulatory strategy* is premised on an impersonal and contractual relationship between state and market. Here, the state's responsibility is to establish policies to strengthen market functions and prevent market failures. Rather than a reduction of state intervention, the regulatory strategy involves a redirecting of activity toward market-enhancing measures. In other words, state intervention is enacted to ensure a fair competitive environment and boost the functioning of price signals. Examples include antimonopoly rules, property rights protection, and standards setting. WTO-specific examples include sanitary and phytosanitary measures (SPS). These rules are market-enhancing in that they provide a common set of rules for firm competition that are evenhanded and transparent rather than allowing individual governments the ability to set their own domestic regulations that might be based on nonscientific or protectionist principles. SPS regulations require that national standards need to be premised on scientific evidence and cannot violate the principle of nondiscrimination.[25] They further require members to notify the WTO of any changes in their domestic SPS requirements and allow other states to scrutinize the scientific basis on which such standards are set. These principles aim to guard against protectionism, and by having all member states adhere to such guidelines, WTO rules on standards arguably enhance the contestability of markets for all members. The regulatory strategy is hence most consistent with WTO principles promoting a noninterventionist and arm's-length role for economic governance.[26]

It should be noted that these state strategies as conceptualized here are distinct ideal types that cannot be arranged on a continuum of high versus low degrees of state intervention. Indeed, *all* of these state strategies involve substantial state intervention, and the difference between each rests on the form rather than quantity of intervention.[27] While some analyses equate greater state activity with a diminished scope for market mechanisms, this book treats states and markets as complements rather than substitutes.[28] For example, the enactment of more regulatory strategies simply implies greater state activism of the market-enhancing type and the strengthening of policies aimed at improving the functioning of the market economy rather than any movement toward the developmental or directive strategies.

Are these strategies mutually exclusive and collectively exhaustive? While framed theoretically as ideal types, these strategies could in reality be deployed in different combinations by Chinese bureaucrats. As chapters 3 through 5 show, policies issued by various actors within the state can—and often do—contain a mixture of strategies. With WTO entry, however, substate actors might shift the relative weight placed on a particular strategy. That is to say, it is possible for an actor to issue a set of policies that are predominantly developmental but contain elements of the regulatory or directive strategies and to shift the relative emphasis of the strategy mix in response to WTO entry. This same mixture can occur even within a single policy document—and indeed, this flexibility is deployed in the machine learning algorithm used to identify the prevalence of various strategies in Chinese industry policies (described in detail in chapter 3). These strategies therefore are less coherent than, say, a military strategy might be. Nevertheless, they represent deliberate strategic choices made by actors within the state to advance their political objectives.

This typology is collectively exhaustive in terms of capturing the main options available to states for pursuing the common goal of economic growth. Indeed, the three strategies are broadly consistent with the dominant economic models found across the world: the planned economies enacted under communism. the developmental states associated with countries such as Japan and Korea, and the market economies common to developed democracies in the West. Sticking to framings premised on protection versus liberalization, in contrast, would subsume both directive and developmental strategies in the "protection" category, thereby conflating two different strategies, both of which involve state intervention but are grounded in distinct state-economy relations.

The three state strategies are defined broadly and intended to complement rather than supplant the rich and growing scholarship on the diversity of ways in which public policies can be designed to effectively spur development. For example, the "midwifery" and "husbandry" state roles proposed by Peter Evans are closely related and would both belong to the developmental market-shaping strategy.[29] Recent work on new industrial policy highlights the many ways in which the state could play a market-shaping role: it could act as a strategic collaborator to the private sector or support the learning and capability development of firms.[30] The state could also play a market-enhancing role by intervening in cases of market failure (e.g., in capturing knowledge spillovers or addressing coordination failures).[31] What distinguishes this analysis from the broader literature on the developmental state and industrial policy is that much of the latter has been oriented toward the question of effectiveness: either explaining cases of or prescribing approaches for success. In contrast, this study is focused on examining strategic responses to globalization and as such is relatively less focused on the important topic of how *effective* these strategies might be.[32]

I also set aside the consideration of state strategies that might be deployed in pursuit of other goals. Examples include policy goals—such as security or education—that might be on the national agenda but are not directly relevant to WTO entry and hence would not be expected to shift as a result of trade liberalization. I further exclude strategies that might be related to trade but are focused on social outcomes rather than economic growth. One example is the compensation policy that has been the subject of many studies on the relationship between globalization and welfare provision.[33] This policy, as its name suggests, is aimed at using government spending to mitigate the negative social effects of globalization such as inequality but is not directly related to the growth imperative outlined above.

Additionally, I do not focus on the important issue of corruption within the state (i.e., predatory state strategies). This is not to say that there might not be ways in which trade liberalization might either curb or encourage corrupt behavior. For example, the transparency and legal requirements embedded in WTO rules might circumscribe opportunities for state predation. On the other hand, WTO-led liberalization might also increase the flow of foreign capital into a country, thereby expanding the opportunities for rent seeking. Nevertheless, the main reason for setting predatory behavior outside the scope of inquiry is that WTO rules themselves are not targeted at changing state behavior related to corruption, whereas the key relationship that I investigate is between WTO membership and various state strategies for growth.

One might also ask if these three state strategies reflect ideological *beliefs* or strategic *choice*. While there is an ideological component to the scope and content of each of the state strategies presented here, I argue that ideologically driven policy responses are more likely to prevail among leading thinkers who advise the government rather than central agencies or the rank-and-file cadres that by and large are driven by political survival within China's bureaucracy.[34] The intense competition to generate economic growth means that most cadres are motivated by pragmatic considerations and are therefore willing to deploy any strategy as long as it promises to boost their political advancement. As for central agencies, an agency's approach to economic governance tends to be largely fixed over time. This is due to China's process of reform described earlier, whereby reform initiatives were implemented alongside the creation of new bureaucratic bodies working around other existing ministries. As such, the specific modes of governance associated with a set of reforms tends to be embedded within individual agencies and to persist over time, as chapter 4 explains in greater detail. While this adherence to a particular set of strategies might arguably be viewed as an ideological attachment, it is not possible to disentangle ideological beliefs from self-interested action, as these agencies have therefore come to see the deployment of a particular set of state strategies as synonymous with the advancement of their bureaucratic interests.

Likelihood of Sanction and Prospects for Political Advancement

Economic governance in China therefore involves contestation between these three state strategies for growth. What explains the adoption of one strategy over another in response to WTO entry? To take into account the vast size of the Chinese party-state and the complex political relationships between the various actors that comprise it, I focus on strategic responses in three dimensions of governance: across administrative levels, between central party-state actors, and across industries. While these dimensions are broad and cross-cutting, this chapter offers a simple overarching explanation that rests on two broad mechanisms: the *likelihood of sanction* and *prospects for political advancement*. Each of these mechanisms affects bureaucratic behavior through a set of different intervening variables depending on which analytical dimension is being studied.

Likelihood of Sanction

WTO membership generates responsibilities on nation-states to bring their domestic economic policies in line with the WTO's articles of agreement. Perceived violations of these WTO rules can be brought to the body's dispute settlement mechanism, thereby generating demands on member governments to account for their domestic economic policies to the international community. While the WTO itself has no formal sanctioning power, it has formalized processes for authorizing retaliation when states fail to comply with dispute panel rulings.

In addition, members keep track of each other's policies on an individual basis, while the consistency of member policies with WTO agreements are discussed multilaterally at various WTO committees. Potentially inconsistent measures that are raised by more than one member in particular can increase the informal pressure on a country to adjust the policy in question and also serve as warnings to individual members about policies that have a higher risk of triggering a complaint at the dispute mechanism. Apart from this member-driven monitoring, the WTO itself publishes periodic trade policy reviews of individual members. Formally, these trade policy reviews cannot be used as a basis for bringing evidence into dispute settlement cases, and member countries are not obliged to respond to the reviews by altering their domestic policies.[35] However, the monitoring and disclosure involved in WTO reviews may nonetheless informally raise external scrutiny over a country's trade and economic governance and in doing so encourage governments to voluntarily make their domestic policies consistent with WTO rules.[36] For China, the transitional review mechanism included in its accession protocol further empowered WTO members to regularly monitor China's implementation performance for a period of ten years, thereby generating another layer of scrutiny over its domestic policies.[37]

This study emphasizes that politics within the state plays an important role in shaping responses to globalization. In particular, large decentralized states can face internal coordination problems in getting their subnational authorities to comply with international obligations.[38] And while China is governed by a single party, internal accountability relationships within the Chinese party-state are highly fragmented and shifting over time, a product of not just China's decentralized administrative structure but also the structure of political networks within its authoritarian regime. Therefore, the likelihood of sanction for violating WTO rules is filtered through China's domestic governance structure and falls unevenly across different bureaucratic actors within the state. This dynamic occurs between not only administrative hierarchies but also the party leadership and the central government. The varying probability of punishment in turn generates different likelihoods that a given actor within the state will respond to WTO entry with a regulatory strategy (which is broadly consistent with WTO rules) or a nonregulatory response (i.e., developmental or directive strategies, which while sometimes allowed under specific circumstances and in specific forms, are broadly inconsistent with the liberal market principles of WTO rules).

Likelihood of Sanction by the WTO

Due to the Chinese bureaucracy's decentralized structure, some parts of the state face a higher likelihood of WTO sanction than others. These costs associated with WTO sanction can broadly be divided into reputation, litigation, and implementation costs. China, it can be argued, faces higher reputational costs from violating WTO rules compared to other member states due to fears over whether its authoritarian regime will be a status quo or revisionist power and the disruptive potential of its economic size. China's behavior within the WTO carried high reputational stakes precisely because WTO membership was seen as a means to bind China more tightly to the rules of the international order. China's record of WTO compliance was therefore intimately tied to the question of whether it would play by the rules. Consequently, its domestic policies came under close scrutiny from WTO members via the transitional review mechanism and from the United States via annual reports by the United States Trade Representative. Indeed, scholars have argued that higher reputational costs explain China's fairly strong record of WTO implementation and relatively high rate of compliance with WTO dispute panel rulings compared to other major economies such as the United States and the European Union.[39] However, these reputational costs fall unevenly within the Chinese state. With decentralization, reputation costs fall primarily on the central government as sovereign representative of the nation to a broader international community, while subnational authorities such as provinces and counties are relatively shielded from the immediate demands of international economic diplomacy.

Litigation costs are related to the administrative and institutional resources required in filing and responding to WTO trade disputes, while implementation costs have to do with the bureaucratic and economic resources involved in carrying out dispute panel decisions. In terms of litigation costs, it is again the central (rather than subnational) government that is responsible for filing or responding to a WTO complaint and which shoulders the burden of implementing the resulting panel decision. Subnational governments, by contrast, do not have to bear the direct costs of any action that might result in a WTO dispute. For example, if the city of Ningbo enacts policies for a particular industry that generate a WTO complaint, it is the central bureaucracy, not the Ningbo city government, that has to undertake the process of representing China at the WTO dispute process. If the complaining state were to retaliate, the form of retaliation might be targeted at a different sector entirely and is highly unlikely to be targeted at Ningbo specifically. This central-subnational dynamic reflects a larger collective action problem: subnational authorities gain most of the benefits arising from the policies they enact, while the reputation and litigation costs of any resulting WTO complaint are externalized to the central government.

As for the costs of implementing and adjusting to WTO rules or dispute panel rulings, it is similarly the central government that is overall in charge of driving the policy process and ensuring implementation. While subnational governments are undoubtedly involved as well, they are largely expected to follow the policy direction set by the central government. In sum, taking into account the combined reputation, litigation, and implementation costs associated with WTO entry, I argue that an actor at a higher administrative level is likely to bear more of the costs of WTO sanctions. Consequently, an actor at a higher level of administration is more likely to adopt a *regulatory* rather than directive or developmental response to the WTO.

Likelihood of Sanction within the Party-State

Why would the central government not be able to directly discipline subnational actors who deviate from WTO rules, such that the costs of WTO sanction are felt at lower administrative levels? While WTO entry has increased the pressure on the central government to ensure national compliance with WTO commitments, the center's ability to do so is hobbled by imperfect monitoring and weak enforcement mechanisms. These weak enforcement powers are a function of several factors, from the sheer size of China and of the CCP to the policy autonomy delegated to subnational levels and the layered governance system where each administrative level is in charge of appointing and managing officials "one level down" (e.g., the prefecture is managed by the provincial government directly above it and not by the central government).[40] From 2002 onward the central government tried to strengthen its supervision and enforcement powers by introducing a new rule that local officials would be held to account for failing to implement central policies. However, as scholars have found, the central government simply does not have the capacity to implement such a system on an institutionalized or sustained basis, leading enforcement actions to be conducted in an ad hoc manner.[41] As a result, subnational violations of central policies and guidelines are commonplace, in recognition of the low probability of being disciplined by the central government.

Accountability relationships further complicate the interests and coordination between actors operating at the very heart of the Chinese party-state—that is, the political leadership and its central government. Given the size of the Chinese bureaucracy, the central government itself is extremely large. One study estimates that in 1998, there were over seven million people working in the central government alone—the size of a small country.[42] It should therefore come as no surprise that there are deep disagreements among major economic agencies over how to govern the economy and that these agencies are often engaged in protracted battles with each other for policy influence. Additionally, interests within the central bureaucracy are not necessarily aligned with those of the political leadership, despite single-party rule. While the central bureaucracy has overall responsibility for implementing China's WTO commitments and other political decisions (and should therefore act as an agent of the party), we cannot assume that the state will always be equally motivated to faithfully carry out leadership goals. Importantly, WTO entry generates winners and losers within the bureaucracy, leading to both support and resistance. Therefore, we need to consider how WTO entry affects the interests of different economic agencies and when the leaders of an authoritarian regime are able to effectively discipline the central government to act as its agent.

As noted earlier, relations between the party leadership and the central government in China's party-state system can be described as one of reciprocal accountability. What, then, determines when the party leadership can effectively deploy the central government as its agent, and when is the state instead a constituent of the leadership?[43] I address this question by assessing the degree to which leadership ties are embedded within the central bureaucracy. While all party leaders have had to work their way up the party bureaucracy before reaching the top, their particular career trajectories vary in important ways. I use the concept of embeddedness to assess the degree to which a party leader's network of support has been built via a career spent primarily rising up through various agencies in the central bureaucracy in Beijing (high embeddedness) rather than rotating through various prefectures and provinces across China (low embeddedness).

When network ties of China's political leadership are *embedded* within the central state, one might expect that leadership era to be associated with a highly effective administration, as the leaders are veterans of the central bureaucracy, and their supporters are drawn from the same set of agencies and organizations. I argue, however, that this network embeddedness acts counterintuitively as a political liability for these leaders, who therefore cannot disentangle themselves from the mutual obligations and ties that have enabled their rise through the bureaucracy. As a result, a party leadership with embedded ties cannot discipline central government agencies for deviating from their delegated tasks without also inflicting that same punishment on members of its own network. Instead, party leaders have to rely on the central state as a key constituent for shoring up their political strength.

In contrast, when the network ties of China's political leadership are *disembedded* from the central bureaucracy, this means that the leaders' network of political support and mutual obligations is in the provinces and outside the Beijing bureaucracy. Consequently, these leaders can effectively punish central government agencies that deviate from their policies without simultaneously inflicting those punishments on members of their own network. Under such circumstances, the party leadership is able to effectively discipline and deploy the central government as its agent for implementing political goals.

This variation in network embeddedness is therefore highly consequential. Depending on where leadership ties are nested, the party-state relationship that dominates (agent versus constituent) affects whether responses to WTO entry are driven by the party leadership or the central government. This distinction affects the resulting policy response, because while the political leadership was generally supportive of regulatory reforms, the central government was split between agencies oriented toward regulatory strategies for growth that stood to gain policy influence from WTO entry and agencies advocating developmental strategies that were hostile to WTO rules that might potentially threaten their policy discretion.

In sum, while new WTO rules on state behavior raised the potential costs of implementing developmental or directive policies at odds with the general market-liberal orientation of the WTO, the degree and direction in which WTO membership would alter state behavior is mediated by certain features of China's domestic institutions. Its decentralized and fragmented internal structure generates uneven likelihoods of sanction at different administrative levels, such that the central government is more likely to bear a disproportionate share of the costs of WTO sanctions while also being unable to shift those costs onto subnational governments due to major deficiencies in its monitoring and enforcement capacities. Additionally, whether pro-WTO leaders in China's authoritarian party-state system are able to advance their liberalizing agenda depends on the embeddedness of these leaders' political networks. Depending on their career trajectory, the location of their networks (rooted in the central bureaucracy versus out in the provinces) affects whether leaders can effectively deploy the central bureaucracy as their agent with a credible threat of punishment or if they have to instead rely on these agencies as a key source of political support. Together, these features of China's decentralized administrative structure and the embeddedness of networks in its party-state system generate varying probabilities of being sanctioned, therefore creating scope for different government actors to adopt divergent (and nonregulatory) responses to the WTO.

Prospects for Political Advancement

Chinese government actors balance the probability of being sanctioned by WTO rules with the equally important calculus of how best to improve their prospects for political advancement, defined broadly as the attainment of greater influence and/or rank within the party-state. Specifically, we need to consider two channels through which international trade rules enhance or undermine a state actor's prospects for political advancement. First are new economic channels introduced by liberalization that affect actors' ability to achieve the economic growth so important to their promotion prospects, and second are bureaucratic channels created by WTO rules that shift the ability of actors to maintain or enhance their policy influence.

Economic Channels

With WTO entry, the range of factors affecting economic growth—and hence the promotional prospects of bureaucratic actors—accordingly shifts toward the threats and opportunities generated by trade liberalization. While in pre-WTO periods officials on average might have adopted economic strategies suited to a more insular environment defined by fairly high tariff barriers, WTO entry alters their strategic calculus to incorporate new open economy factors affecting growth. These factors range from how industrial diversification affects the relative prospects of export promotion versus import competition to the potential gains to be extracted from external resources such as FDI and foreign technology and how best to secure state assets and maximize employment in a globalized economy.

While the likelihood of WTO sanction affects whether actors across administrative levels adopt regulatory strategies in response to WTO entry, I propose that the choice between nonregulatory strategies (i.e., developmental or directive) depends on how liberalized economic competition affects the growth prospects of different administrative authorities. China's entry into the WTO is often seen as primarily an opportunity to expand Chinese exports.[44] This view, however, underplays the degree to which WTO entry equally lowered tariffs on Chinese imports, greatly increasing external competition to China's domestic industries. Trade liberalization brings both export opportunities and import competition threats, and much of the negotiating around China's WTO accession was centered on the reduction of China's import tariffs to expand market access for foreign firms. China's trading partners were able to negotiate substantial reductions, with manufacturing tariffs cut to an average of 9.4 percent and those for agriculture to 17.4 percent.[45] Much of the bureaucracy's anxiety and resistance surrounding WTO entry had to do with the threat of cheaper foreign imports undermining China's less competitive industries, from automobiles to agricultural products. However, this anxiety was far from uniform. Whether different government officials saw WTO entry as presenting a source of export-oriented gains or import-driven losses depends on the underlying competitiveness of their jurisdictions.

I propose that for different jurisdictions, the balance between whether WTO entry posed largely an export opportunity versus an import threat rests on the diversity of that unit's industrial base, that is, the degree to which the economy within a jurisdiction comprises a wide or narrow range of different industries. Industry diversity shapes a government's response to WTO entry by affecting the relative prospects of export promotion versus import competition resulting from trade liberalization. On average, officials in charge of a highly diversified industrial base are likely to find that the risk of WTO-induced import competition is fairly *diffuse*, as any one industry tends to make up a relatively smaller proportion of the economy. A more diversified economic base further means that there is a higher likelihood that some of the industries in the jurisdiction will find that they have a comparative advantage in producing under open economy conditions. As a result, officials overseeing highly diversified jurisdictions are more likely to see WTO entry as providing greater export opportunities relative to import competition and enact more *developmental* rather than directive policies to actively harness the potential benefits of trade liberalization to generate growth.

In contrast, officials overseeing a jurisdiction with a poorly diversified industrial base are likely to find that the risks of WTO-induced import competition are more *concentrated*, as even a few negatively affected industries can make up a large share of the economy. As a result, officials in such jurisdictions are likely to view WTO entry primarily as an economic threat associated with job losses and a decline in industrial activity. These officials are therefore more likely to enact *directive* policies to forcefully intervene and mitigate the economic dislocations associated with import competition.

All other factors held equal, governments at higher administrative levels are likely to have a larger jurisdiction size and a more diversified industrial base (as will be shown in chapter 3). Lower-level jurisdictions such as China's numerous counties and prefectures tend on average to be more poorly diversified due to their smaller size and often disadvantaged geographical location in western or inland China. Hence, I hypothesize that governments at higher administrative levels—when they are not pushed toward a regulatory strategy by the threat of sanctions—are more likely to adopt a more developmental response to the WTO, while those at lower levels are more likely to adopt a directive response.

Another factor to consider in the economic channel is the expansion of FDI flows into China that resulted from WTO entry and how this increase in foreign capital affected competition for political advancement within the state. Here, I focus on how political advancement generates divergent incentives over the promotion of strategic industries. Importantly, central and subnational governments pursue different types of political advancement. I argue that the central government as a whole (including the party leadership) is deeply concerned with the continued political durability of the CCP. Therefore, the central government treats economic growth as a means to *regime promotion* and focuses on raising the long-term viability of the economy by emphasizing value-added growth and the promotion of state assets in sectors considered to be strategic lifelines of the economy. In contrast, the subnational government is led by party secretaries competing with each other to be promoted up the next rung of the party hierarchy. These subnational leaders therefore treat economic growth as a means to *rank promotion* and seek to maximize short-term growth rates to boost their chances of career advancement.[46] In contrast to the goals of the center, the subnational government seeks to maximize economic output and employment in the immediate term.

While these center-subnational tensions have long been a feature of China's political economy, WTO-led liberalization enhances access to external resources such as FDI that exacerbate the divergence between central and subnational goals. I argue that while both central and subnational governments seek to attract FDI into China, their differing focus on regime versus rank promotion leads them to deploy this FDI toward contrasting political purposes. With WTO entry, the central government seeks regime promotion by maximizing access to the foreign technologies embedded in FDI, particularly for the promotion of strategic sectors where state assets are concentrated. Consequently, the central government focuses its developmental policies on raising the technology content of strategic industries. In contrast, WTO entry provokes subnational governments to pursue their rank promotion imperative of maximizing short-run growth by emphasizing the importance of the foreign capital embedded in FDI. Therefore, in the WTO context, subnational states are likely to adopt developmental policies aimed at raising foreign capital inflows. This divergence in the purposes to which FDI is deployed—technological advancement versus capital investment—ends up confounding central policies, because subnational strategies aimed at maximizing short-term capital do not generate business conditions suitable to technological upgrading. As a result, central policies aimed at promoting national champions in the WTO era end up being riven by greater internal contradictions.

Bureaucratic Channels

Bureaucratic actors seek not only to be promoted up administrative ranks but also to enhance their influence over the economic agenda. This is particularly true when we examine the central government agencies in charge of implementing China's WTO commitments, which are locked in heated contests for influence and control over the direction of economic policy.[47] WTO rules that specify economic governance principles that the state should abide by go far beyond potentially constraining—or not constraining—domestic policy space, as much literature has focused on. Rather, these rules alter the prospects of central agencies for expanding their bureaucratic influence. While some agencies might find their discretion circumscribed and their scope for expanding their policy influence much more restricted under the WTO, the same rules might provide external leverage to other agencies to advance their policy agenda, providing them with more opportunities for political advancement.

I propose that in general, technocratic agencies in charge of regulatory issues such as standards setting are likely to support WTO-led reform, as the implementation of specific WTO rules oriented toward building a regulatory state would enhance their policy relevance and influence.[48] They are therefore likely to respond to WTO entry with intensified regulatory strategies aimed at expanding their reach over economic governance. In contrast, agencies in charge of setting industrial policy are likely to be more developmental in orientation and hence likely to be opposed to WTO rules that threaten to restrict the set of policy tools at their disposal. They are therefore likely to seek opportunities to respond to WTO entry with developmental strategies aimed at defending and expanding their policy authority.

What determines which of these sets of agencies will prevail in the contest for influence? My conjecture is that whether agencies oriented toward a particular state strategy are able to advance their agenda depends on the strength of leverage to be obtained from WTO rules, which varies over time. China's protocol of accession to the WTO specified a detailed timeline for the implementation of different commitments. The pressure for implementing WTO rules would therefore be strongest in the years immediately following WTO entry and gradually decline over time as various commitments listed in the protocol schedule are enacted. Therefore, agencies supportive of regulatory reform are most likely to be able to use China's WTO commitments as leverage to push for greater influence over the reform agenda in the period immediately following WTO entry. As these commitments get fulfilled, the strength of WTO leverage weakens over time. This then puts agencies supportive of developmental strategies in a potentially stronger position to advance their policy agenda.

In sum, China's authoritarian state is riven by competing political interests despite single-party rule: across administrative levels, between the party leadership and the bureaucracy, and between central agencies. How various political-bureaucratic actors respond to the new conditions introduced by WTO entry depend first on their varying probability of being sanctioned. Second, their WTO policy responses also rest on the ways in which their prospects for political advancement shifted under WTO entry.

Predicting Strategic Responses to WTO Entry

Figure 2.1 summarizes the key features of my argument. WTO entry reconfigured the alignment of "winners" and "losers" within China's fragmented state via an economic channel by introducing newly liberalized conditions for growth and via a bureaucratic channel by imposing new rules for conducting economic policy. The strategic responses of various Chinese government actors to these new conditions in turn depend on how the bureaucratic and economic aspects of WTO entry affect their probability of being sanctioned and

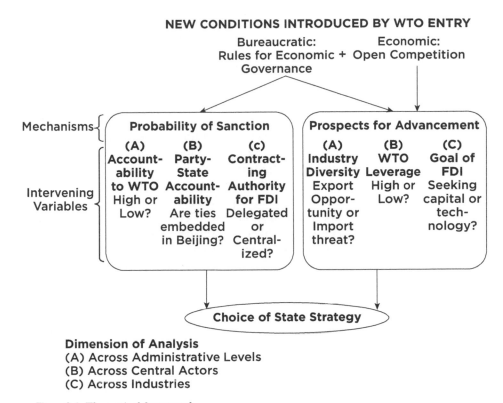

Figure 2.1. Theoretical framework

prospects for political advancement. Given the nature of domestic institutions in China, which combine fragmented accountability relations with intense political competition, the exact nature of sanction and advancement shifts depending on which dimension of the complex party-state is being analyzed. The following section summarizes the predicted strategies that different government actors are likely to adopt in responding to WTO entry. I present my hypotheses as they relate to three dimensions of analysis: across administrative levels, across central actors, and across industries.

Strategic Responses across Administrative Levels

Due to China's decentralized structure, an actor's *degree of accountability to WTO rules* varies depending on where it sits within the administrative hierarchy. Governments at a higher administrative level, being more accountable to the demands of international economic diplomacy, face a stronger likelihood of sanction for violating WTO rules and are therefore more likely to adopt regulatory responses to WTO entry. Governments at lower administrative levels, being more insulated from WTO rules, face a lower likelihood of sanction and are therefore more likely to adopt directive or developmental responses to the WTO. At the same time, these subnational authorities are directly exposed to the economic effects of WTO liberalization and are highly motivated to respond to defend growth rates in their jurisdictions. The choice between these nonregulatory strategies then depends on the *industry diversity* of a given administrative unit, as this diversity affects the economic implications of WTO entry (see intervening variables A in figure 2.1). Governments at a higher administrative level are more likely to govern a jurisdiction with a more diversified industrial base, for which trade liberalization is likely to bring greater export opportunities relative to import competition. Such actors are therefore more likely to adopt developmental responses to WTO entry. Governments at a lower administrative level are more likely to oversee a jurisdiction with a less diversified industrial base that is relatively more vulnerable to import competition and less well positioned to exploit export opportunities brought by trade liberalization. Consequently, these lower-level governments are more likely to adopt directive rather than developmental responses to WTO entry.

Given these dynamics, the central government, which faces the highest likelihood of sanction and oversees the most diversified industrial base, could adopt either regulatory or developmental responses to the WTO. The provincial government, which is one level removed from WTO sanctions but still oversees on average jurisdictions with relatively diversified industrial bases, is more likely to adopt a largely developmental response to the WTO. The local government (e.g., counties and cities) has the lowest likelihood of facing sanctions from the WTO and on average oversees jurisdictions that are poorly diversified and is therefore more likely to adopt a largely directive response to WTO entry.

Strategic Responses across Central Government Agencies

While China's political leaders were largely pro-WTO in their outlook, the strategic responses emanating from the central government shifted over time. I argue that this change in central policies is the product of shifts in the party-state accountability relationship and the declining leverage of WTO rules (see intervening variables B in figure 2.1). Whether the political leadership can effectively deploy its central bureaucracy as an agent to effect regulatory reforms in line with WTO rules is contingent upon the location of leadership networks. When a party leader's political network lies largely outside the central bureaucracy as a result of a career spent in the provinces, the leader can effectively punish central agencies without simultaneously inflicting the punishment on members of his network. Under such circumstances, central agencies face a high probability of sanction for deviating from the leadership's pro-

WTO preferences and are hence more likely to adopt regulatory strategies in response to WTO entry. When a party leader's political network is largely embedded *within* the central bureaucracy as a result of a career spent largely rising through central agencies, the leader cannot punish the bureaucracy for policy deviations without also punishing members of his own network. Rather than deploying the state as its agent, the leadership in this case needs to rely on the state as a core political constituent. As a result, central agencies face a low likelihood of sanction by the leadership for adopting nonregulatory strategies. Under such circumstances, economic policy would then largely be driven by central agencies rather than the party leadership.

When policy responses are primarily driven by central agencies, the question of which type of strategy prevails depends on the extent to which WTO rules can be used as leverage in the interagency competition for influence. Agencies supportive of regulatory reform are likely to benefit from WTO rules, as these rules potentially enhance their standing within the bureaucracy. Hence, they have an interest in deploying China's WTO commitments as leverage to push for greater influence over the reform agenda. This strategy is more likely to be effective in the years immediately following WTO entry, when the pressure for implementing WTO rules was strongest due to a clear timetable for implementation laid out in China's accession protocol. As the strength of WTO leverage weakens over time, however, rival agencies supportive of developmental strategies are likely to be in a stronger position to push for their policy agenda.

As a result, the central government is more likely to enact regulatory responses to WTO entry during periods of time when party leadership networks are disembedded from the central bureaucracy (allowing the pro-WTO leadership to effectively deploy the state as its agent) and in the early years of WTO entry when WTO leverage was still strong (empowering regulatory agencies to advance a more liberalizing agenda). Conversely, the central government is more likely to enact developmental responses to WTO entry during periods when party leadership networks are deeply embedded within the central state (when leaders need to rely on state bureaucrats as a political constituency, thereby allowing the bureaucracy to drive the policy agenda) and in the later years of WTO entry when WTO leverage had weakened (enabling developmental agencies to influence the policy trajectory).

Strategic Responses across Industries

As this chapter has emphasized, the central government's ability to discipline subnational authorities is hobbled by imperfect monitoring and weak enforcement mechanisms, allowing subnational governments to deploy strategies aimed at advancing their own political interests. This generates divergence in the policies that central and subnational authorities adopt across different industries, a tension that is exacerbated by globalization due to the surge in FDI flows into China after WTO entry. When contracting authority over FDI is delegated to subnational governments, the center is thus unable to enforce its policy priorities, to deleterious effects for its efforts to promote national champion industries. This contracting authority is consequential, because while both actors seek to attract FDI, their distinct interests mean that the goal for which FDI is deployed is at times conflicting (see intervening variables C in figure 2.1). The central government, motivated by a long-term regime promotion imperative, seeks FDI to achieve technological upgrading in the economy and is therefore likely to adopt more developmental strategies to boost technology transfer. The subnational government, motivated by a short-term rank promotion imperative, seeks FDI to maximize capital for immediate output and employment growth. It is therefore likely to adopt more developmental strategies to maximize access to foreign capital.

The theory presented in this chapter provides an explanation for why WTO entry did not result in a monolithic convergence toward liberalization. It does so by focusing the spotlight on China's fragmented domestic state structure and explicating the channels through which WTO entry affects the incentives of substate actors in the bureaucracy. While China's top leadership and its foreign trading partners sought to use WTO entry as a credible commitment to drive domestic reform, the resulting policy trajectories that emerged cannot be explained through existing frameworks premised upon outcomes such as constraint or compliance. These outcomes are largely dichotomous in nature and therefore obscure the broader range of policy responses emerging from the Chinese state. By conceptualizing the range of potential responses in terms of three competing state strategies—market substituting, market shaping and market enhancing—I highlight the multifaceted ways in which actors in the Chinese state have responded to new conditions introduced by WTO membership.

By disaggregating the Chinese state into its various political actors that affect economic policymaking, I show that the punishments and opportunities associated with WTO entry do not fall evenly within the single-party state. Rather, various government actors respond strategically and differentially to WTO rules to advance their political interests. In general terms, the observed variation in policy trajectories result from a combination of a government actor's likelihood of being sanctioned and its prospects for political advancement in the context of WTO-led liberalization. These two broad variables, while set up in a simple cost-and-benefit framework, manifest themselves through different intervening channels depending on which part of the massive party-state is being studied. The result is a divergence of policy responses to a set of external rules that was designed to bind China more tightly to the international system.

The next three chapters test the hypotheses presented here, drawing on quantitative and qualitative methods that will be explained in detail in each chapter. Chapter 3 focuses on the hierarchical division of power within China to explain why we observe varying strategic responses to WTO entry across administrative levels, while chapter 4 looks within the powerful central government to explain a shift in central policy trajectories over time, from one of market-enhancing liberalization to one of market-shaping developmentalism. Finally, chapter 5 explains why WTO entry has undermined rather than strengthened the governance of China's national champion industries.

Chapter 3

The Hierarchical Politics of WTO Entry

There is an old saying in China that "the mountains are high, and the emperor is far away." This phrase, which obviously predates communist rule, speaks to a perennial challenge of governing a country of massive geography, population, and diverse regional socioeconomic conditions. The central authority, no matter how powerful, confronts major monitoring and enforcement problems vis-à-vis its subordinate units. Knowing this, China's trading partners were careful to address the country's internally heterogenous and decentralized governance structure when setting down the terms for China's accession to the World Trade Organization (WTO). Section 2 of the accession protocol targets the need for uniform administration of China's trading regime, committing the government to implement its WTO commitments across all subnational levels of government as well as regions that would have enjoyed preferential treatment such as special economic zones, development zones, and industrial parks. In addition, China agreed to set up a complaint mechanism whereby inconsistencies in domestic trade-related regulations could be reported to the central government.

That China's trading partners included these clauses into the protocol reveals their concerns over the subnational reach of the WTO agreement. Indeed, data from a 2005 American Chamber of Commerce survey of US firms located in China reveals sharp differences in business assessments of central versus subnational government implementation of WTO rules. Whereas 20.5 percent of respondents thought that the central government was "willing, able, and prepared to implement changes in the spirit of the WTO agreement," only 10.5 percent thought that subnational governments held the same attitude. Conversely, 21 percent of respondents thought that subnational governments were "actively seeking loopholes in the requirements to avoid or delay implementation," while 11 percent thought that the central government held the same attitude. Why did foreign firms have such pessimistic assessments of subnational government behavior, and what drives these differences in WTO responses across administrative levels?

This chapter focuses on the divergent incentives faced by the central, provincial, and local (i.e., subprovincial) governments that make up China's hierarchical bureaucracy. Within China's authoritarian structure, these subnational authorities are powerful entities in their own right. In terms of geographic size and population, individual provinces are comparable to large countries around the world. The province of Guangdong, for example, has thirty million more people than all of Germany.[1] As much literature has pointed out, provinces occupy an influential intermediary position in China's governance structure. Formally, provinces possess the authority to exercise substantial policy discretion because they hold the same rank as central ministries and are therefore not bound by central-level directives. Provincial leaders often deploy this authority in pursuit of interests that may not be aligned with those of the center. Informally, provinces are further able to exploit gaps in the center's monitoring and enforcement capabilities to ignore or even subvert central priorities. These dynamics remain salient and in some ways have even become reinforced in the Xi Jinping era as power has become increasingly concentrated at higher levels, giving provincial authorities a stronger gatekeeping role in the allocation of important economic resources.[2]

In contrast to provinces, local governments—prefectures, counties, and townships—sitting at lower rungs of the administrative hierarchy tend to be much smaller in size and population. Just like their provincial superiors, they are driven by their local conditions to pursue policies that may not align with those of the central government. Sitting far away from national-level politics in Beijing, these local authorities are by and large shielded from China's international economic diplomacy at the WTO. And while some of the richer counties and cities on the coast enjoy vibrant economies, the vast majority of these local governments tend to oversee jurisdictions with relatively circumscribed resource endowments compared to provinces. As this chapter will show, these two factors—a government's degree of accountability to the WTO and the diversity of its industrial base—drive varying strategic responses to the new economic and bureaucratic conditions introduced by WTO entry.

Degree of Accountability to the WTO

With China's entry into the WTO, the Chinese bureaucracy came under new pressure to conduct its economic policies in ways consistent with international trading rules. Violations of these rules can be challenged formally through the WTO dispute settlement system and informally through pressure exerted at WTO committee meetings as well as the China-specific transitional review mechanism. As discussed in chapter 2, the sanctions associated with violating WTO rules come in different forms, from reputation to litigation and implementation costs. Broadly speaking, the threat of sanction raises the probability of a governmental actor choosing a market-enhancing policy response, as such regulatory strategies are most consistent with WTO principles advocating an arm's-length relationship between state and market. However, the likelihood of sanction for violating WTO rules varies depending on where one sits in China's administrative hierarchy.

For large decentralized countries, the responsibility of representing one's domestic policies at the international level falls primarily on the central government as the sovereign representative of the nation. This means that the central government is likely to be more sensitive to the costs of potentially WTO-inconsistent policies compared to subnational authorities. Unlike the national government in Beijing, provinces and localities are relatively sheltered from the immediate demands of international economic diplomacy and are not directly responsible for dealing with a WTO dispute. A collective action problem ensues whereby the benefits of developmental or directive policies enacted at subnational levels accrue directly to the subnational jurisdiction, but the potential costs of WTO sanction are borne by the central government.

The central government could in theory use its enforcement powers to ensure that subnational governments also abide by WTO rules. This would reduce the likelihood of complaints or questions being raised by members through the various WTO mechanisms discussed earlier. However, central authorities often have to deal with the fact that from the perspective of many local governments, "the mountains are high, and the emperor is far away." Entrenched monitoring and enforcement challenges discussed in the previous chapter mean that Beijing is often unable to discipline subnational authorities who deviate from national policy. As a result, subnational defiance of central policies is a common occurrence, despite the center's efforts to strengthen its internal enforcement mechanisms.[3] These factors—the central government's position as sovereign representative at international levels and the relatively sheltered position of subnational units—combine to mean that governments at higher administrative levels are likely to face a higher threat of WTO sanction and thus are more likely to adopt a regulatory rather than nonregulatory (directive or developmental) response to the WTO.

Diversity of the Industrial Base

A government's choice between a directive or developmental response is in turn affected by the pursuit of political advancement that permeates China's political system, as chapter 2 discussed. While subnational leaders are commonly under pressure to demonstrate positive growth results in their jurisdiction, their ability to do so is directly affected by the much more liberalized economic environment and intensified foreign competition brought by WTO entry. Some leaders would find their own jurisdictions to be more well positioned to benefit from WTO-led liberalization than others. I propose that one important factor affecting the economic impact of WTO entry on a jurisdiction is the diversity of a unit's industrial base. By industry diversity, I refer to the degree to which the economy of a jurisdiction comprises a wide or narrow range of different industries. Industry diversity shapes a government's response to WTO entry by affecting the likely risks of import competition relative to export opportunities resulting from trade liberalization.[4]

On average, officials in charge of a highly diversified industrial base are likely to find the risks of WTO-induced import competition to be fairly diffuse, as any individual industry that might face competition from cheaper imports tends to make up a smaller proportion of the economy. A more diversified economic base further means a higher likelihood that some of the industries in the jurisdiction will discover a comparative advantage in producing under open economy conditions and expand their exports. On balance, officials overseeing more highly diversified jurisdictions are therefore more likely to see WTO entry as providing export opportunities. They are then more likely to enact *developmental* policies based on market-shaping measures to harness the potential growth benefits of trade liberalization, thereby boosting their prospects for political advancement. In contrast, officials overseeing a jurisdiction with a poorly diversified industrial base are likely to find their economy more vulnerable to the threat of WTO-induced import competition, as even a few negatively affected industries can make up a large share of economic activity. As a result, officials in such jurisdictions are likely to view WTO entry primarily as an economic threat associated with job losses and declines in industrial activity. These officials are therefore more likely to enact *directive*, market-substituting policies to more forcefully intervene and mitigate the economic dislocations associated with import competition.

On average, authorities at higher administrative levels, such as provinces, are likely to have a larger jurisdiction size and a more diversified industrial base compared to lower-level counties and townships.[5] This means that a provincial government is more likely to see WTO entry as a source of export opportunities and adopt a developmental strategy as a response. Local authorities, in contrast, being smaller and less industrially diversified on average, are more likely to see WTO entry as bringing a higher risk of import competition relative to export opportunities. They are therefore more likely to adopt a directive, market-substituting strategy in response to this threat.[6]

As the central government's reputational, litigation, and implementation costs of violating its international obligations are all higher relative to subnational authorities, the central government arguably faces a higher threat of sanction from violation of WTO rules. It is therefore more likely to respond to WTO entry with regulatory policies. At the same time, given that the central government's jurisdiction is over the entire country, its policies are also informed by governance over a highly diversified industrial base. From Beijing's perspective, the threat of import competition resulting from WTO entry is likely to be fairly diffuse compared to export opportunities, leading the central government to be more likely to adopt a developmental response to the WTO. Which of these two competing pressures will dominate the central government's response? I pose this question as an empirical test of the strength of WTO rules on the Chinese central government. A dominant regulatory response would indicate that WTO rules have a strong disciplinary effect in driving central government policies toward a regulatory strategy. If the effect of WTO rules were weaker, WTO entry would possibly induce a strengthening of both regulatory and developmental policies, or a largely developmental response.

Provincial governments do not have to directly account for their actions at the international level. Nonetheless, provincial leaders do hold a high rank equal to that of central ministers and participate in central-level state and party bodies such as the National People's Congress and the Central Committee. Provincial authorities would therefore be involved in passing new legislation and policy decisions enacted to comply with China's WTO obligations. Given that provincial leaders are competing with each other for promotion to higher positions, they could also be expected to be more responsive to the preferences and concerns of the central government relative to local authorities. Therefore, the province's likelihood of sanction by WTO rules can be said to lie between the central and local government, rendering the province more likely than local authorities to adopt a regulatory response to WTO entry. In terms of industry diversity, provinces are on average much larger than local governments and hence are likely to oversee a more diversified industrial base. As such, while the share of industries threatened by WTO-induced import competition will be somewhat larger at the province relative to the central government, the industrial base of the average province is likely to be still diversified enough to be able to absorb the impacts of import competition while allowing its government to pursue export-driven growth by picking and choosing winners across different industries. Provincial governments are therefore more likely to respond to WTO entry with developmental rather than directive policies. The province's relatively more indirect accountability to WTO rules and disciplines also means that pressures to enact regulatory strategies will tend to be weaker compared to the central government. Indeed, since the province is one level removed from WTO sanctions and yet enjoys a highly diverse industrial base, its developmental response to WTO entry is likely to be stronger than any regulatory response.

Local governments are characterized by weak accountability to WTO rules. Not only are local leaders far removed from international economic diplomacy related to the WTO, but they also report to the provincial or subprovincial governments directly above their jurisdiction rather than the central government. This weak accountability means a low likelihood of facing direct sanctions from violating WTO rules and hence fairly diffuse pressure on local authorities to respond to the WTO's regulatory demands. The smaller size of local jurisdictions further implies that on average, local governments also have a less diversified industrial base and are likely to face more concentrated losses from WTO-induced import competition compared to provinces or the central government. These forces combine to suggest that local governments are on average likely to focus their response to WTO entry with directive measures aimed at mitigating the negative effects of trade liberalization.

Using Textual Analysis to Measure State Strategies

One major difficulty of this study involves coming up with a measure that captures the changing enactment of regulatory, developmental, and directive state strategies: across industries, across different parts of the state (central agencies and subnational governments), and over time. While some studies have documented this internal contestation and major debates over economic governance within China, to the author's knowledge no one has yet tried to assess the competing strategies quantitatively or tested the various factors that might affect the adoption of different economic governance strategies by different governmental actors.[7]

This book captures the flux and heterogeneity in the bureaucracy's deployment of competing state strategies using textual analysis of an original data set of Chinese industry policy documents. The language in Chinese government regulations, policies, and measures is revealing of different strategies because—as discussed in chapter 1—although China's leadership has long established economic growth as a political priority, the question of what measures the state should then adopt to achieve this growth is a subject of persistent disagreement within the party-state. These disagreements find their way into policy documents even at the national level.

For example, in central economic guiding documents such as the Five-Year Plan and other policy announcements, the words used to describe state-market relations have changed over time but also reflect enduring tensions between the market-enhancing strategy and other types of governance approaches. The Tenth Five-Year Plan (2001–2005) noted that "market mechanisms are playing, increasingly markedly, a *basic role* in the distribution of resources."[8] Yet in December 2001 (the same month of China's WTO accession), then Vice Premier Wu Bangguo noted that China needed to foster fifty globally competitive state-owned enterprises (SOEs) as part of "an important strategy to cope with the impacts brought about by the country's accession to the WTO."[9]

The Eleventh Five-Year Plan (2006–2010) stated that China was "giving *more play* to the fundamental role of the market in allocating resources *under guidance of macro regulation and control.*"[10] At the same time, the plan also introduced a focus on "indigenous innovation," and in 2009 a new policy priority was placed on "strategic emerging industries," heightening state activism in technology innovation and in industries such as new-energy vehicles, high-end equipment manufacturing, biotechnology, and information technology. Also in 2006, the head of the State-owned Assets Supervision and Administration Commission of the State Council (the agency in charge of managing central SOE assets) announced that the state would retain "absolute control" over seven core "strategic sectors" and that central SOEs should establish themselves as "heavyweights" in six other sectors.[11] These mixed messages continued into the following decade. While the twelfth plan (2011–2015) gave "*full play to the socialist mechanism as well as to the market* in terms of allocating resources," a 2013 Central Committee communiqué further noted that the market should play a "*decisive role*" in resource allocation.[12]

These conflicting statements reflect the disagreements within the party-state (discussed in chapter 1) over the appropriate measures of economic governance. This study's approach is therefore to look within the bureaucratic texts of industry regulations for language reflecting competing state strategies for growth. Indeed, these industry policies and measures can potentially tell us more about a government's preferred economic strategy than common industrial policy measures such as subsidies or taxes. Evidence of high levels of subsidies, for example, does not reveal whether a governmental actor is trying to protect a declining industry or engage in developmental policies for a nascent sector, thereby potentially conflating two different types of strategies, both of which involve substantial state intervention but are driven by distinct conceptions of the state's role in the economy. In addition, economic data is arguably a more accurate measure of policy *outcomes*, which can be the result of conflicting policy approaches put forth by different actors, as well as distortions introduced through the process of policy implementation. By contrast, the words found in policy documents are a more accurate measure of the outcomes of interest in this study, which are the modalities underpinning the relationship between state and market in each of the three state strategies for growth.

The corpus of documents collected includes all publicly available policies related to China's manufacturing industries from 1978 to 2014, issued by subnational governments as well as the full array of central agencies involved in economic governance. This is a departure from existing studies of industrial policies, which tend to focus on major national or sectoral policies issued by a few central bodies that lay out the overarching policy objectives and targets for industrial development, such as the State Council, the National Development and Reform Commission (NDRC) and the Ministry of Industry and Information Technology (MIIT)). To the extent that local-level deviation from central policy is examined, this is typically done through in-depth case studies of select cities or industries.[13] I take a more encompassing approach of capturing the policy content of all players large and small, central and subnational, involved in governing an industry. This inclusive approach is important, because some agencies not formally in charge of industrial policy for a sector can nonetheless have a substantial impact on its development. In the automotive industry, for example, formal policy has been the purview of the State Council, the NDRC, or the MIIT. However, the Ministry of Transportation is also involved in automotive governance in terms of regulating the number of cars permitted on the road, while the Ministry of Public Security oversees traffic safety and the Ministry of Environmental Protection sets vehicle emissions standards. National policies are also often broad enough that subnational governments in turn issue their own local-level implementing policies that may deviate from the intentions of central agencies. The corpus of documents analyzed in this study therefore captures much fuller variation in policy content for the governance of an industry both within the central government and across different administrative levels, compared to existing analyses of Chinese industrial policies.

As the data collected is industry- and location-specific, I am further able to combine the textual data with available data on industry and province-level characteristics. This allows us to not only have a consistent measure of industrial policy content over space and time within different parts of the Chinese state but also quantitatively test the relationship between changes in policy content and changes in industry and regional socioeconomic conditions.

Data Collection and Measurement of State Strategies

The corpus of documents was collected by taking each of China's manufacturing industries as listed in the four-digit China Industry Code and searching for policies related to that industry in an online repository of Chinese laws and regulations.[14] The search results showing policies, regulations, and measures related to each industry were then scraped from the website, with the final data set comprising 43,069 documents across 128 manufacturing sectors ranging from food processing to textiles, steel, aircraft, medical equipment manufacturing and so on. In the data, 14,831 documents are issued by central government agencies and bodies, 18,162 by provincial governments, and 10,076 by local authorities (subprovincial prefectures and counties).

The documents collected are industry-specific in that only those policies and measures for which the industry name appears in the title were collected. Higher-level policies, such as Five-Year Plans, and national-level industrial policies covering multiple sectors are not included in this data set. However, we can expect priorities in national-level guidelines to be reflected in these industry-specific policies, as China's top-down policy process means that major goals announced in national plans are later concretized in industry-specific policies by individual agencies and lower-level governments. The collected documents range from legislation approved by the National People's Congress to State Council and central ministry regulations, legal interpretations issued by the Supreme People's Court, and legislation, guidelines, and regulations issued by subnational governments. The substantive issues contained in these documents span many topics, such as taxation, standards supervision, labor management, certification, financial supervision, environmental protection, enterprise management, government procurement, public safety, customs affairs, and more. Table 3.1 provides a selection of document titles showing the range of topics that tend to be covered by various governmental bodies.

To what degree do these documents reveal genuine policy intent as opposed to cheap talk or political signaling? That is, could the language found in these policy documents be strategically chosen by issuing agencies or governments to signal their obeisance of central leadership priorities? As table 3.1 shows, the documents in the corpus tend to be focused on highly specific issues—investment approval for an individual firm or standards-setting for a particular product. As such, most of these documents belong to the mundane life of an economic bureaucracy rather than high-profile political documents aimed at catching the eye of higher-level leaders that might therefore be more likely to be strategically phrased with a view toward signaling policy fealty. Moreover, if the documents collected largely reflected strategic signaling, they would all likely contain *similar* language reflecting the leadership priority of the day. Instead, as the following sections will show, there is substantial variation in the language contained in the documents—even for those governing a common industry—issued by different parts of the Chinese state as well as over time. In order to explain these differences in policy language, it is important to look beyond a strategic signaling explanation and consider the factors driving the choice of particular economic strategies by various government actors.

Due to the large size of the corpus, I use a machine-learning algorithm to uncover the differences in policy language across the thousands of industry documents. This algorithm (called the correlated topic model) generates clusters of words—or topics—that are likely to belong together based on their co-occurrence across the corpus of documents.[15] It also assesses the likelihood of each document belonging to a particular topic as well as the correlation between different topics based on the co-occurrence of the topics. Since the corpus is made up of industry regulations and measures, the documents are all implicitly focused on the challenge of industry governance—and the broad national objective of economic growth. However, the topics generated (forty-one in total) reveal the many different issues and approaches involved in governing the economy. The topics range from a focus on trade and antidumping to accounting, permits, development, science and technology, legality, and rectification. To validate the substantive focus in each of the topics, each topic was given a descriptive label after a careful reading of the top twenty words and the top ten documents most likely to belong to the topic.

The next step of the analysis involves identifying the topics most representative of the directive, developmental, and regulatory strategies. Not all of the topics generated would be relevant to—or reflective of—these state strategies, and testing all forty-one topics would also raise the likelihood of false positives. I therefore take the more conservative approach of selecting only those topics most directly reflective of the outcome of interest (strategies) for analysis. This process resulted in two topics identified as being representative of each state strategy, or a total of six topics.[16]

Table 3.1 Sample of document titles

Level	Document Title	Year
Local	Circular by the Shenzhen Municipal People's Government on Issuing Measures to Promote the Development of the Shenzhen Semiconductor LED Industry	2009
Provincial	Circular by the Shanghai Foreign Trade and Economic Cooperation Commission's Foreign Trade Planning and Operation Department on the Issuance of a Temporary Export License for Type 5 Textiles Exported to the European Union	2005
Central	Circular by the State Tobacco Monopoly Bureau on Strengthening the Management of the Tobacco Monopoly and Tobacco Agency in Beijing	1997
Central	China Securities Regulatory Commission Approval of the Exemption for Dalian State-Owned Assets Investment Management Group Ltd. of Obligations in the Tender Offer for Dalian Rubber & Plastics Machinery Co., Ltd. Shares	2011
Central	State Council General Office Forwarding a Circular by the State Economic Commission, State Planning Commission and Ministry of Light Industry on the Consultation over a Few Questions on the Development of the Salt Industry	1986
Local	Circular by the Weichang Manchu and Mongol Autonomous County People's Government Office Endorsing the County Environmental Protection Bureau and Supervision Bureau "Opinion on the Strengthening of Pollution Prevention and Control Work in the Potato Starch Processing Industry"	2007
Central	State Council Circular Approving the Opinion by the State Planning Commission and the Ministry of Railways on the Development of Central and Local Joint-Venture Railways	1992
Provincial	Approval by the Jiangsu Province State Taxation Bureau on Preferential Tax Benefits for Taizhou Weidmann High Voltage Insulation Co., Ltd.	2004
Local	Shenzhen Special Economic Zone Regulations on the Metal Material Trading Market [not in effect]	1994
Central	Ministry of Finance Opinion on Several Issues regarding State-Owned Assets Management in the Tobacco Industry	2004
Central	State Administration of Quality Supervision, Inspection and Quarantine Circular 2003 No. 17: Verification of the Application by Yantai Wine to Use Geographical Origin Product Labeling	2003
Central	Ministry of Railway on Deepening the Reform of the Railway Industry Science and Technology System [not in effect]	1989
Central	Circular by the Ministry of Finance on the Tariff Policy for the Importation of Pesticides (Medicine and Original Drug)	2003
Provincial	Gansu Provincial Department of Environmental Protection Approval of the Application by Pingliang Conch Cement Co., Ltd. for Emission Permits	2014

(*continued*)

Table 3.1 (continued)

Level	Document Title	Year
Central	China Food and Drug Administration Bulletin 2013 No. 34: Circular of the No. 1 Amendment to YY0334-2002 "General Requirements for Silicone Rubber Surgical Implants" and two Medical Device Industry Standards	2013
Provincial	Beijing Municipal Pricing Bureau and Beijing Municipal Finance Bureau Approval of the License Plate Fee for the "92" Type of Tractor	1995
Central	Ministry of Labor and Social Security Letter on Wages Owed by China Heavy Vehicles Enterprise Group and Other Issues	1999
Central	Letter from the Intellectual Property Division of the Supreme People's Court on the Dispute Case regarding Xuzhou Optical Instrument Factory, Xuzhou Medical Optics Electronic Instrument Research Institute, and Li Guoqiang's Infringement of Copyright and Trade Secrets	1999
Local	Tongchuan Municipal People's Government on Further Accelerating the Development of the Ceramic Industry	2008
Central	Circular by the Ministry of Health Pharmaceutical Affairs Bureau on Issuing National Standards (GB) Numbers for Biological Products	1991
Local	Circular by the Office of the Hechi Municipal People's Government on the Establishment of the Leading Small Group for the Hechi City Non-ferrous Metals High-Tech Industrial Park	2005

A reading of the documents with a high probability of belonging to the directive strategy found that these policies and regulations tend to be concerned with state intervention oriented toward reorganizing industry via a reliance on traditional tools of administrative guidance. The first directive topic is labeled "Rectification," and documents in this topic frequently contain terms such as "control," "shut down," and "put in order." Documents with a high probability of belonging to the second directive topic ("Crackdown") similarly reflect state intrusion, with frequently occurring terms such as "punish," "rectify and improve" and "realign." Figure 3.1 shows an example of a directive document, a Dalian city regulation giving instructions for the closure of unlicensed mines and the consolidation of small unprofitable mines. As can be seen in the figure, directive policies frequently employ the use of highly interventionist terms such as "rectify," "shut down" and "restructure/correction," reflecting the reliance on government measures rather than market mechanisms to shape industrial outcomes.

In English, an excerpt of the document reads:

Circular by the Dalian Municipality People's Government General Office on Issuing the Work Plan for the **Rectification** and **Shutting Down** of Metal and

Figure 3.1. Example of a directive document

Non-Metal Mines in Dalian from 2013 to 2015 (Da Zheng Ban Fa [2013] No. 56) Work plan for the **rectification** and **shutting down** of metal and nonmetal mines in Dalian in 2013–2015. This plan is established to thoroughly implement the spirit of the "Circular by the Liaoning Province People's Government General Office on Issuing the Work Plan for the **Rectification** and **Shutting Down** of Metal and Non-metal Mines in Liaoning Province in 2012-2015" (Liao Zhengban [2012] No. 69), to effectively undertake the work of **rectifying** and **shutting down** metal and nonmetal mines and to ensure that the work of **rectification** and **shutting down** is effective, in combination with realistic conditions in the city.

1. Guiding Ideology and Working Principles

(I) Guiding Ideology

Under the guidance of the Scientific Outlook on Development, we have thoroughly implemented the spirit of the Party's 18th National Congress, and strengthened the safety production infrastructure of noncoal mining enterprises, in accordance with the holistic planning of the Liaoning Province People's Government General Office Circular on the Provincial Work Plan on the **rectification** and **shutting down** of noncoal mines issued by the Provincial Economic and In-formatization Commission and other departments (Liao Zheng Office [2012] No. 1), to further refine the scope, objectives, procedures, and standards as well as to clarify the work objectives of the **rectification** and **shutting down** of mines so as to fully complete our city's work task of **rectifying** and **shutting down** mines.

(II) Working Principles

Under the unified leadership of the municipal government, all relevant depart-ments shall jointly enforce the law and, in accordance with the strategy of mov-ing northward and under the principles of eliminating backwardness, treating both symptoms and problems and steady advancement, adopt overall measures such as **shutting down**, reorganization, **restructuring and correction**, and im-provement to replace or **shut down** unlicensed mines and small mines that do not have safe production conditions in accordance with the law to comprehen-sively increase the production safety level and safety capability of the mines.

In contrast, a reading of documents with a high probability of belonging to the developmental topics found that these policies and regulations tend to be exhortatory in tone and explicitly discuss industry promotion. The first devel-opmental topic is labeled "technology upgrading," and documents found in this topic frequently use words such as "accelerate," "encourage," "innovation," and "technology." The second topic is labeled "processing development," and documents in this topic use words such as "develop," "market," "scale," and "build," mostly focused on supporting the development of lower value-added industries. Figure 3.2 shows an example of a document in the developmental category, a Nanjing city policy to vigorously develop and stimulate innovation in the semiconductor industry that frequently employs use of the terms "de-velop," "science and technology," and "high-speed" or "speed up."

In English, an excerpt of the document reads:

Nanjing Municipal Government Office forwards the Circular from the Munici-pal Economic Commission on "Implementing Opinion on **Speeding Up** the City-wide **Development** of the Semiconductor LED Industry."

The people's governments of all districts and counties, municipal bureaus and commissions, as well as directly subordinate units: The "Implementing Opinion on **Speeding Up** the City-wide **Development** of the Semiconductor LED Indus-try" formulated by the Municipal Economic Commission has been approved by the municipal government and is now forwarded to you. Please implement it conscientiously.

December 12, 2004, Implementing Opinion on **Speeding Up** the City-wide **De-velopment** of the Semiconductor LED Industry (Municipal Economic Com-mission, November 2004). The semiconductor LED industry is another emerg-ing industry in the electronic information industry. It has the characteristics of

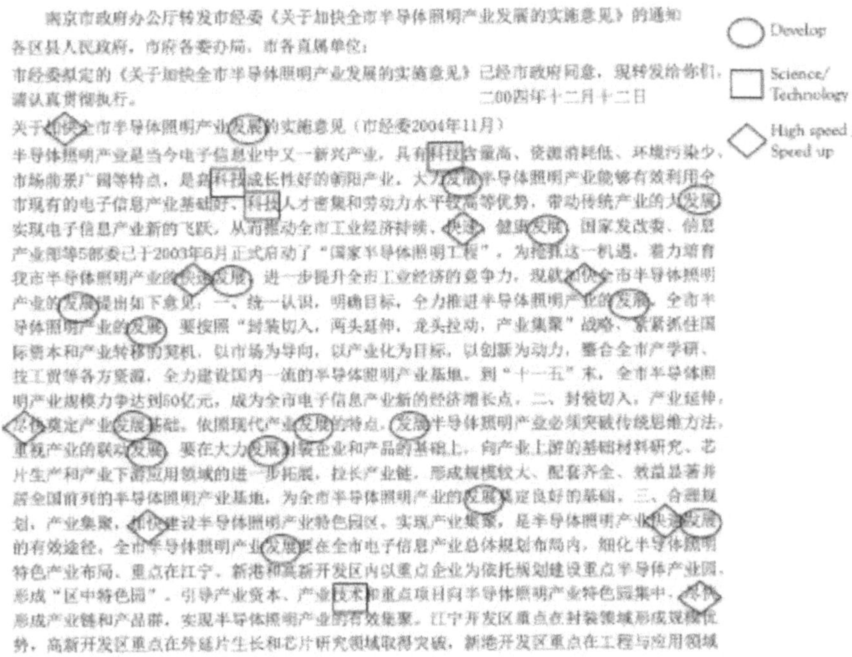

Figure 3.2. Example of a developmental document

having a high technology content, low energy consumption, less environmentally polluting, and broad market prospects. It is a sunrise industry with good high-tech growth. By vigorously **developing** the semiconductor LED industry, we can effectively use the city's existing good foundation in the electronic information industry, intensive scientific and technological talents, high productivity levels, and other advantages to drive the **development** of traditional industries and achieve a new leap in the electronic information industry, thus contributing to the sustainable, **high-speed**, and healthy **development** of the entire municipal industrial economy. The National Development and Reform Commission, the Ministry of Information Industry and other five ministries officially launched the National Semiconductor LED Project in June 2003. In order to seize this opportunity, to put effort into cultivating the city's **high-speed development** of the semiconductor LED industry, and enhance the competitiveness of the city's industrial economy, we will now **speed up** the **development** of the city's semiconductor LED industry by putting forward the following opinions: First, with a unified understanding and a clear goal, to spare no effort in promoting the **development** of the semiconductor LED industry.

A reading of the documents classified as belonging to the regulatory strategy found that these policies tend to be neutral and technocratic in tone, reflecting

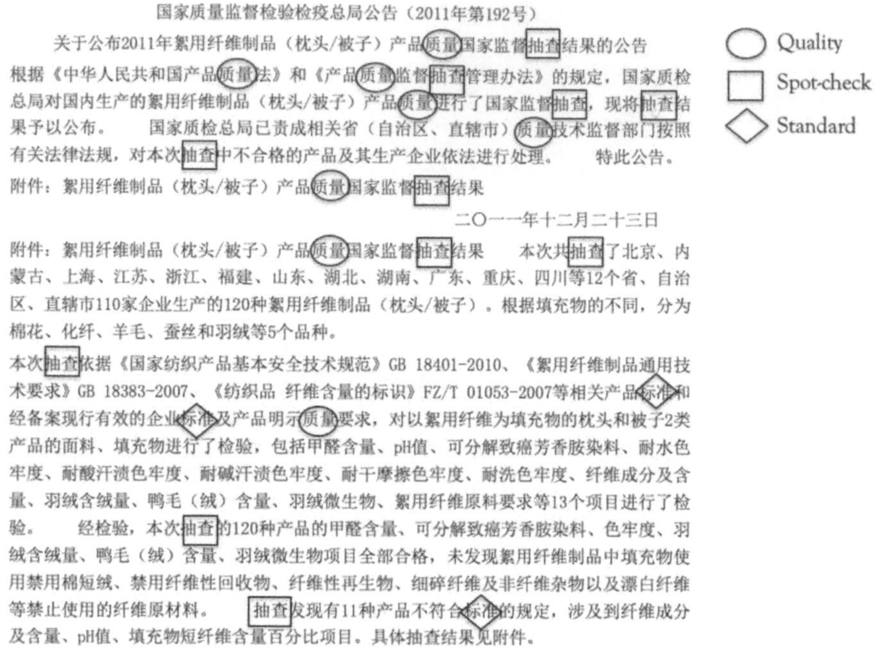

国家质量监督检验检疫总局公告（2011年第192号）

关于公布2011年絮用纤维制品（枕头/被子）产品质量国家监督抽查结果的公告

根据《中华人民共和国产品质量法》和《产品质量监督抽查管理办法》的规定，国家质检总局对国内生产的絮用纤维制品（枕头/被子）产品质量进行了国家监督抽查，现将抽查结果予以公布。　国家质检总局已责成相关省（自治区、直辖市）质量技术监督部门按照有关法律法规，对本次抽查中不合格的产品及其生产企业依法进行处理。　特此公告。

附件：絮用纤维制品（枕头/被子）产品质量国家监督抽查结果

○ Quality

□ Spot-check

◇ Standard

二〇一一年十二月二十三日

附件：絮用纤维制品（枕头/被子）产品质量国家监督抽查结果　本次共抽查了北京、内蒙古、上海、江苏、浙江、福建、山东、湖北、湖南、广东、重庆、四川等12个省、自治区、直辖市110家企业生产的120种絮用纤维制品（枕头/被子）。根据填充物的不同，分为棉花、化纤、羊毛、蚕丝和羽绒等5个品种。

本次抽查依据《国家纺织产品基本安全技术规范》GB 18401-2010、《絮用纤维制品通用技术要求》GB 18383-2007、《纺织品 纤维含量的标识》FZ/T 01053-2007等相关产品标准和经备案现行有效的企业标准及产品明示质量要求，对以絮用纤维为填充物的枕头和被子2类产品的面料、填充物进行了检验，包括甲醛含量、pH值、可分解致癌芳香胺染料、耐水色牢度、耐酸汗渍色牢度、耐碱汗渍色牢度、耐干摩擦色牢度、耐洗色牢度、纤维成分及含量、羽绒含绒量、鸭毛（绒）含量、羽绒微生物、絮用纤维原料要求等13个项目进行了检验。　经检验，本次抽查的120种产品的甲醛含量、可分解致癌芳香胺染料、色牢度、羽绒含绒量、鸭毛（绒）含量、羽绒微生物项目全部合格，未发现絮用纤维制品中填充物使用禁用棉短绒、禁用纤维性回收物、纤维性再生物、细碎纤维及非纤维杂物以及漂白纤维等禁止使用的纤维原材料。　抽查发现有11种产品不符合标准的规定，涉及到纤维成分及含量、pH值、填充物短纤维含量百分比项目。具体抽查结果见附件。

Figure 3.3. Example of a regulatory document

an arm's-length approach to governance aimed at enabling, rather than substituting or shaping, market functions. The first topic is labeled "Standards," and documents belonging to this topic tend to use terms such as "inspect," "spot-check," and "quality testing." Documents in the second topic ("Certification") similarly contain regulatory words such as "certify," "label," and "standard." Figure 3.3 shows an example of a regulatory document, a central government announcement of the results of a spot-check of product standards in the pillows and blankets industry that frequently uses the words "quality," "spot-check," and "standard."

In English, an excerpt of the document reads:

Announcement of the General Administration of Quality Supervision, Inspection and Quarantine (No. 192 of 2011)

Announcement on Promulgating the Results of State Supervision and **Spot-Check** on the **Quality** of Fiber Wadding Products (Pillow/Quilt) Products in 2011. Pursuant to the "Product **Quality** Law of the People's Republic of China" and the "Administrative Measures on Product **Quality** Supervision and **Spot Checking**," the State Administration of Quality, Supervision, Inspection and Quarantine carried out national supervision and **spot-checks** on the **quality** of domestically produced fiber products (pillows/quilts) and will now announce the results. The SAQSIQ has instructed the relevant provincial (including autonomous regions and province-level cities) quality and technical supervision departments to deal with the **unqualified** products and their manufacturing enterprises found in this **spot-check** in accordance with relevant laws and regulations.

Hereby announced. Attachment: Results of national supervision and **spot-check** of product **quality** of fiber wadding products (pillow/quilt). December 23, 2011.

Attachment: The results of the national supervision and **spot-check** on the **quality** of products made from fibrous products (pillows/quilts) include 120 kinds of fibrous products (pillows/quilts) produced by 110 companies in 12 provinces, autonomous regions and province-level cities: Beijing, Inner Mongolia, Shanghai, Jiangsu, Zhejiang, Fujian, Shandong, Hubei, Hunan, Guangdong, Chongqing and Sichuan. The products were divided into five varieties based on their different fillers: cotton, chemical fiber, wool, silk and down. The **spot-checks** were conducted according to relevant national product **standards** such as the "National Basic Safety Technical Specifications for Textile Products" GB 18401-2010, the "General Technical Requirements for Fiber-based Products" GB 18383-2007, the "Identification of Textile Fiber Content" FZ/T 01053-2007 and other current corporate **standards** and product **quality** requirements.

The measurement of the three state strategies is operationalized by summing up the prevalence of the two topics associated with each strategy. With this measure, we can then assess changes in the prevalence of each strategy in industry regulations and measures issued by governments at different administrative levels and over time. Topic prevalence is a meaningful measure of relative policy importance in the Chinese context because when it comes to official documents, "the more an issue is discussed or a word is used, the more important it is."[17]

One question that might be raised about the regulatory strategy is whether the documents classified in this category are really market-enhancing in nature, given that regulatory measures might be enacted by governments as a form of hidden protectionism or nontariff barriers. To test this possibility, figure 3.4 shows the difference in strategy prevalence in policy documents issued by coastal versus inland subnational governments. Coastal governments, with their strategic location and history of being promoted as special trading zones, can be expected to be more highly competitive in their trade activities and consequently more market-oriented in nature. In contrast, inland localities, with their more sheltered economies, are likely to be more vulnerable to import competition and hence potentially more protectionist. Therefore, if the topics classified in the regulatory category are actually protectionist in substance, they should be more prevalent in policy documents issued by inland rather than coastal authorities. Figure 3.4 shows that the opposite is the case. Subnational governments located along the coast are *more* likely to adopt regulatory language in their industry policies, while inland governments are more likely to adopt directive and developmental language. This figure also highlights the diversity

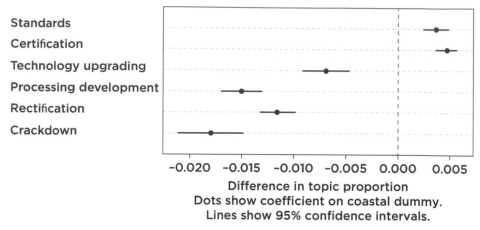

Figure 3.4. Difference in strategies between coastal and inland subnational governments

in political economic conditions *within* each level of administration. While the theory presented here focuses on average responses across different levels of government, it does not deny that there will be variation between governments within provincial and local levels. Indeed, a later section exploits this very cross-provincial diversity to test the proposed causal mechanisms.

Figure 3.5 show trends over time in the relative prevalence of each strategy at central, provincial, and local levels. The figure shows a distinct divergence in policy trajectories of the central, provincial, and local governments in the years after China joined the WTO. Panel A shows that whereas there was no discernible trend in the years prior to 2001, policy language reflective of a regulatory strategy became most prevalent in central level policies in the WTO years, followed by provincial and then local regulations. In contrast, panel B shows that policy language reflective of a developmental strategy strengthened significantly in the WTO years at the provincial and local levels. Finally, the trends shown in panel C indicate that language reflective of a directive strategy increased in both local and provincial levels in the WTO years but much more strongly at the local level. These descriptive trends provide initial support for the argument that WTO entry has affected the policy trajectories of central, provincial, and local governments differentially, as seen in the divergence of policy trends in the WTO years.

Measuring WTO Entry

I use changes in the tariff levels applied on China's manufacturing industry imports from the 1998–2007 period to assess each industry's exposure to the new economic and bureaucratic conditions introduced by WTO entry. WTO-driven

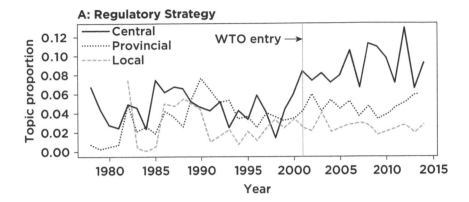

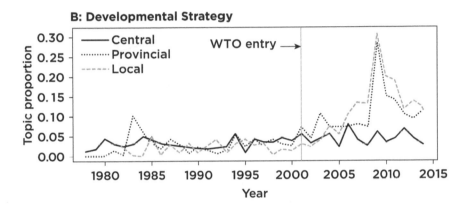

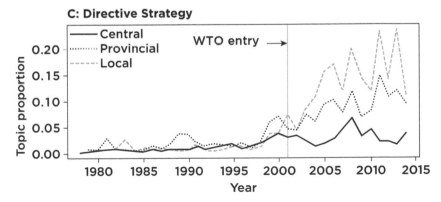

Figure 3.5. Prevalence of regulatory, developmental, and directive strategies in central, provincial, and local regulations
Source: Author's data set.

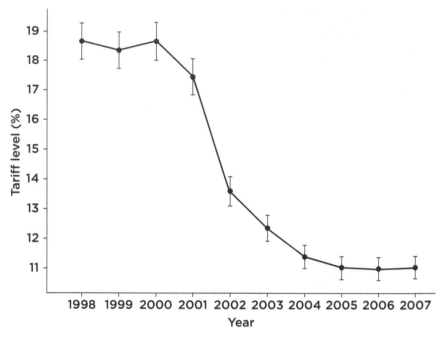

Figure 3.6. Average tariffs imposed on Chinese manufacturing imports
Source: World Integrated Trade Solution.

tariff reductions do not just lower border taxes but also come with an embedded set of rules on conducting trade. This means that industries with lower tariff barriers are more exposed to international trade rules compared to industries whose products are nontraded. Government agencies therefore have to modify their administrative processes and regulations for import-intensive industries to enable compliance with WTO rules on, for example, food and product safety standards. To the extent that lower tariffs also reduce the costs of imported inputs, some industries may find themselves newly competitive on world markets and begin to export. This then raises obligations on state authorities to ensure that the policies surrounding these new exports comply with WTO rules in order to avoid complaints being filed or safeguards being imposed by trading partners. Therefore, changes in tariff levels represent a relevant measure of not just the degree of economic liberalization but also an industry's degree of exposure to trade rules. And indeed, as shown in figure 3.6, WTO entry brought large reductions in the tariffs imposed on foreign imports into China, with most of the cuts occurring in a concentrated period between 2001 and 2006. Average tariffs on manufacturing goods were cut to about 10 percent. The reductions were large even for products whose tariffs ended up higher than that average level. For automobiles, for example, the tax rates were lowered from around 50–60 percent (depending on the specific product) to 25 percent.

As discussed in chapter 1, this negotiated tariff schedule was determined primarily by China's political leadership and its foreign trading partners over the vigorous resistance of a central bureaucracy that would later be in charge of the implementation process. Subnational governments were also not consulted in China's WTO negotiation processes.[18] However, questions remain over whether the pattern of WTO tariff reduction might have additionally been influenced by domestic policy trends and interests such that the size of tariff cuts reflects not just the depth of WTO-led liberalization but also the strength of these domestic factors.

I run two tests to address these issues, the details of which are presented in appendix A. To assess if sectors that faced larger tariff cuts were already on a different policy trajectory compared to sectors for which tariff reductions were smaller, I estimate the relationship between the size of the WTO tariff reduction and the pre-WTO trend of state strategies across industries. To test the possibility that the pattern of tariff reduction might be driven by domestic industry interests, I estimate the relationship between Chinese industry characteristics—such as the industry share of exports and the share of SOE output—in the pre-WTO years (1998–2001) and the size of WTO tariff reductions. The results of these tests offer assurance that that the pattern of China's WTO tariff reductions is by and large not a function of domestic industry pressure but instead is imposed top-down by China's leaders and shaped externally by the commercial interests of China's major trading partners.

Assessing Strategic Responses to WTO Entry

What kinds of state strategies did central, provincial, and local governments adopt in response to WTO entry? The quantitative analysis includes control variables to account for alternative factors that could affect the variation in strategic responses to WTO entry. These include the value of industry exports to account for the possibility that export intensity might directly encourage the adoption of regulatory strategies. Governments might also adopt more developmental or directive strategies for industries that occupy an important position in the economy. For example, more directive strategies might be enacted for high-employment or very large industries, given the potential negative social consequences of unemployment or a fall in output in these sectors. I therefore also include data on employment and industry output in the analysis. Strategic responses might further be affected by firm ownership. For example, we might observe a stronger adoption of directive strategies for sectors dominated by SOEs, while foreign enterprises might lobby for greater market-oriented liberalization in their industries, resulting in more regulatory strategies. Therefore, I also include variables capturing industry output shares of SOEs and foreign and nonmainland enterprises.[19] The main explanatory variable is industry tariffs. For ease of interpretability, this measure is inverted

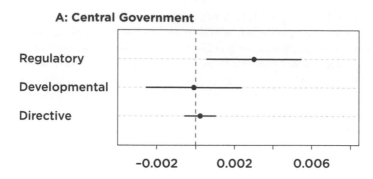

A: Central Government

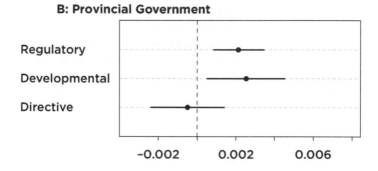

B: Provincial Government

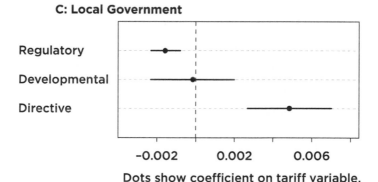

C: Local Government

Dots show coefficient on tariff variable.
Lines show 95% confidence intervals.

Figure 3.7. Topic responses of central, provincial, and local governments to tariff liberalization

such that a positive coefficient on the tariff variable indicates an increase in a particular strategy in response to WTO entry.

The results of this analysis are summarized in panels A to C in figure 3.7, showing the different strategic responses adopted by China's central, provincial, and local governments in response to WTO tariff liberalization. Statistically significant responses do not intersect with the vertical dashed line, while insignificant responses do. A positive value in the dot chart indicates that on average, tariff liberalization for a given industry was accompanied by an increase in the prevalence of that strategy in Chinese industrial policies and measures, while a negative value indicates a decline in strategy prevalence in response to tariff cuts.[20]

The results demonstrate a clear divergence in the responses adopted by central, provincial, and local governments to WTO entry, reinforcing the overall trends shown earlier in figure 3.5. Panel A in figure 3.7 shows that the central government responded to WTO entry with a strengthening of its regulatory, market-enhancing strategies. Substantively, the average tariff cut induced by WTO entry for a given industry led the central government to increase its adoption of regulatory strategies by about 45 percent compared to pre-WTO levels. Notably, the same tariff cut did not induce a change in Beijing's developmental strategy, suggesting that accountability to the WTO outweighed any developmental response that might have been driven by high industry diversity at the central level.

The central government's policy agenda in the years immediately following WTO accession support these statistical results. This period was marked by a legal and regulatory overhaul as central authorities sought to bring their institutions and policies in line with WTO commitments. Institutionally, separate agencies that had overlapping authorities over standards setting and product inspection were merged to form the ministerial-level State Administration of Quality Supervision, Inspection and Quarantine (SAQSIQ). A center was established under the SAQSIQ and Ministry of Commerce to liaise directly with WTO committees on the technical barriers to trade and the sanitary and phytosanitary measures regarding issues related to technical regulations and standards. In the legal arena, amendments were made to the Product Quality Law to strengthen product standards and enhance the enforcement of regulations against substandard or counterfeit goods. The Import and Export Commodity Inspection Law was revised to bring foreign and domestic products under a common certification scheme. The SAQSIQ led a national effort to harmonize Chinese standards with those set by international bodies. An assessment of over twenty-one thousand technical standards by the government led to a revision of over 40 percent and an abolition of about 7 percent of these standards.[21] Official recognition of the need for domestic standards to come into international conformity and the processes for adopting international standards in accordance with WTO principles were further set out in a national regulation issued in December 2001.[22]

In contrast, provincial governments responded to WTO entry with increased developmental *and* regulatory strategies, as panel B in figure 3.7 shows. Consistent with our expectations, the developmental response is much larger compared to the regulatory one. Substantively, the average WTO-induced tariff cut led provincial governments to strengthen their developmental strategies by about 72 percent compared to pre-WTO means. In contrast, provinces on average strengthened their regulatory strategies by about 39 percent—not only far weaker than the developmental response but also of a smaller magnitude than the central government's regulatory strengthening. This increase in regulatory language could reflect provinces' involvement in legislative efforts to comply with WTO rules, generating some accountability to the WTO that is stronger than local governments but weaker than the center.

China's sheer geographical size means that the average provincial government oversees a highly diversified industrial base, and from the perspective of a provincial leader, the risks of WTO entry bringing concentrated import competition is likely to be weaker than the opportunities for enhanced exports. Provinces are therefore more likely to adopt a developmental response to WTO entry oriented toward spurring export-oriented growth. For example, Anhui province issued an industrial policy for its automotive sector in the same month of China's WTO entry that exemplified this developmental approach. Anhui's policy noted that while membership in the WTO would remove many existing protective barriers for China's automotive industry as well as introduce new foreign competition, accession to the WTO would nonetheless reduce input costs, expand exports, and enhance China's integration with the global automotive industry. The document moreover explicitly embraced a market-shaping role for the government by referring to the adoption of industrial policies by other developing countries such as South Korea and urging industrial development through the "combined influence" of the market and government.[23] Similarly, Shandong province issued a developmental policy for its shipbuilding industry in 2002, noting the need to "seize the opportunity" presented by WTO entry to build Shandong into a major shipbuilding province. The policy particularly emphasized the need to expand exports and employ market-shaping measures such as credit and subsidy allocations to support provincial shipping exports.[24]

Panel C in figure 3.7, showing the responses of local governments, indicates yet another set of responses from China's numerous prefectures and counties. Unlike their central and provincial counterparts, local governments strengthened their directive strategies in response to WTO entry, alongside a *weakening* of their regulatory strategies. Substantively, local governments responded to the average WTO-induced decline in tariffs with a *doubling* of directive language and about a one-third decrease in regulatory language in their industry policies. The average local government faces a fairly weak accountability to the WTO because it is the central government that is primarily responsible for representing and defending China's economic policies to the WTO and also because central authorities have long faced challenges with being able to discipline subnational authorities on a consistent basis. Since it is the central government that has to represent and defend China's economic policies to WTO members, the average local authority does not need to take into account any negative spillovers from its policies on the international stage. As a result, pressure on the average local government to adopt WTO-consistent regulatory policies is weak, as these authorities face a fairly low probability of being sanctioned by complaints filed at the WTO.

At the same time, the average local government responded to WTO entry with a surge in directive, market-replacing strategies. This local-level response might seem surprising, given that some of China's most dynamic market-oriented engines of growth are located at the local level. The counterintuitive aspect of this finding stems from the fact that the analysis captures the average response across all local-level governments in China. The directive response of the average local government to WTO entry reflects the fact that most localities in China tend to be more rural and lie inland, with less diversified industrial bases than their dynamic coastal counterparts that we tend to associate with local Chinese economies. As figure 3.4 showed, the strategies adopted by inland localities are indeed more directive compared to their coastal counterparts. These local economies, as chapter 1 noted, operated largely under a separate trade regime sheltered from foreign trade competition and dominated by administrative guidance up until WTO entry. From the perspective of the average prefecture or county leader, WTO entry therefore poses a higher risk of concentrated import competition relative to export opportunities, provoking localities to respond to restructure their uncompetitive industries with direct intervention.

For example, Yanbian prefecture in Jilin province issued a policy in 2003 aimed at consolidating its cement industry by shutting down its small-scale and uncompetitive plants. The market-replacing approach to this restructuring was reflected in the policy's declaration that all plants below scale that had not been shut down by the announced deadline would be "forcibly closed" and have their business license revoked by the local authorities. Banks would cease to issue loans to these small plants, the electricity supply to these plants would be cut off, and production equipment would be "dismantled and destroyed on site."[25] Also in 2003, the Xi'an city government issued a notice to restructure its paper-processing industry by phasing out small-scale and inefficient enterprises. Indicative of a directive approach, the policy set hard production targets, noting that within two years the production scale of Xi'an's paper-processing enterprises had to reach thirty-four thousand tons per year. Smaller-scale plants had to either shut down or merge into larger enterprises.[26]

In short, China's entry into the WTO triggered not one but instead multiple strategic responses across different parts of its hierarchical administrative structure. While we could observe an overall strengthening of regulatory strategies consistent with WTO rules, this response derived primarily from the central government. This market-enhancing response was weaker at the provincial level of government and even declined when it came to local authorities. Instead, provincial governments adopted a strengthened developmental response to WTO entry to pursue export-oriented growth opportunities created by WTO entry, while local governments intensified their directive strategies to mitigate the dislocating effects of import competition in their jurisdictions.

Could these results be driven by a functional separation of responsibilities across administrative levels? According to this alternative explanation, the observed divergence in WTO responses reflects the assignment of standards-setting authority to the central government, while provincial governments have greater influence over policies to develop the economy. This dynamic is unlikely, as standards-setting authority—like many other policy issues—is decentralized to subnational authorities.[27] It might also be that the divergence in strategic responses reflects a tacit coordination of behavior between the central and subnational governments, where the central government, under more intense scrutiny by WTO members, strengthens its regulatory policies but discreetly allows its subnational governments to deviate from national guidelines and subvert WTO rules. To test the theory presented in this study against these alternative explanations, the next section leverages variation across provincial socioeconomic characteristics to test the accountability and industry diversity mechanisms.

The Role of Accountability and Industry Diversity

This section focuses on how the mechanisms of accountability and industry diversity affect the province's developmental strategies, given that provinces exhibited a largely developmental response to the WTO. When it comes to accountability, it is in the interests of the central government to discipline provinces for enacting developmental policies because the center bears a disproportionate burden for managing any WTO disputes arising from subnational actions. However, the likelihood that a province will curb its developmental policies depends on its degree of accountability to the central government.

To come up with a measure of center-province accountability, I utilize a data set created by Bulman and Jaros that captures the years that different members of provincial party standing committees (PPSCs) have spent working in the central government versus the provinces.[28] The PPSC is an important political organ to scrutinize, as it holds the strongest decision-making power within the provincial government. In this data set, a member is coded as having in-province experience if he or she has worked in that province for at least ten years, while a member is coded as having central experience if he or she has worked in Beijing for at least five years. We would expect, for example, that a provincial government whose standing committee in a given year is composed largely of cadres with dominantly in-province work experience would be less accountable to (and behave with more autonomy from) the central government, driven by stronger relationships with interest groups located within the province and/or a deeper understanding of the economic circumstances unique to the province. I measure this province autonomy using the share of committee members who have ten or more years of in-province work experience. Appendix A shows the summary statistics of this measure. As autonomy and accountability are inversely related, I then invert this measure such that a higher value implies greater accountability to the center (i.e., lower province autonomy).

Using this measure, I can then test whether a province's accountability to the central government mediates its developmental response to trade liberalization. As the rationale for provinces adopting developmental responses to the WTO is based on the trade opportunities they expect to gain from their highly diversified industrial bases, I use an interaction term between *province accountability* and a *trade openness* measure (provincial exports plus imports as a share of GDP) as the main explanatory variable in the regression analysis.

I further include a number of variables in the analysis to address alternative drivers of the provincial adoption of developmental strategies. These include provincial GDP, log of GDP per capita, and unemployment levels, as provincial governments overseeing jurisdictions that are less prosperous or have higher levels of unemployment might deploy more developmental strategies to spur economic growth. I also include provincial fiscal balances, as the capacity to adopt developmental strategies (which involve measures such as tax and credit incentives) might be correlated with the health of government finances. In addition, SOEs and even foreign enterprises might lobby their provincial governments to adopt more developmental policies aimed at supporting their sectors. I therefore include the share of foreign direct investment and the share of state-controlled enterprise firms in provincial GDP in the analysis. Finally, I add a measure for industry diversity (described in the next section), as this variable is part of the theoretical framework.[29]

One additional consideration is that the party secretary, as the top provincial leader, might wield disproportionate influence within a PPSC. However, it is unclear how much more weight the party secretary should be given relative to the rest of the committee. To assess how much the results change depending on party secretary influence, I run a series of estimations with weights ranging from 0.1 (highly dispersed influence) to 0.9 (close to monopolistic influence) placed on the party secretary's work experience vis-à-vis the rest of the PPSC.

If stronger accountability to the central government restrains provinces from pursuing their developmental response to WTO entry, we should observe that the developmental response to trade liberalization declines as provincial accountability increases (in other words, a negative coefficient on the *accountability-trade openness* variable). Figure 3.8 shows the plot of the interaction effect when the weight of the party secretary is set at a fairly modest level of 0.2. The figure

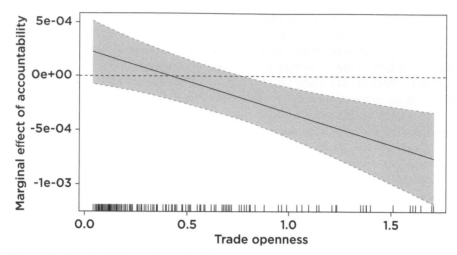

Figure 3.8. Marginal effects of accountability on provincial developmental strategy as trade openness increases
Note: Shaded areas indicate 95% confidence intervals.

shows that developmental strategies decline when a given province is more accountable to the central government and that this decline is larger when a province is more open to trade (see appendix A for supporting tables). The analysis further reveals that changing the weights placed on the importance of the party secretary's work experience does not alter the results. The negative relationship between provincial accountability and developmental responses is statistically significant for the full range of weights placed on the party secretary.

Provinces also vary in their economic structures, and I exploit these interprovincial differences to test the argument that jurisdictions with a more diversified industrial base tend to adopt stronger developmental responses to WTO entry. To do this, I construct an *industry diversity* measure using a province-level Herfindahl index that captures whether a province's industrial output is dominated by a few industries (poorly diversified) or evenly spread out across many industries (highly diversified). (See appendix A for the summary statistics for this variable.)

As the choice between developmental and directive strategies rests on the expected intensity of import competition, I measure the degree to which provinces are exposed to import competition by constructing a *WTO exposure* variable whereby the import tariffs that a province faces are weighted by each industry's share in provincial output. This means that if a particular industry occupies a larger share of a provincial economy, the tariff cuts for that industry will be given a larger weight. The measure is then inverted such that low tariff levels represent a high WTO exposure (i.e., low tariff barriers to trade).

In the analysis estimating the effects of industry diversity in mediating the province's developmental WTO response, the key explanatory variable is there-

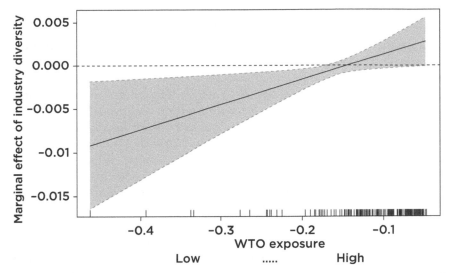

Figure 3.9. Marginal effects of industry diversity on provincial developmental strategy as WTO exposure increases
Note: Shaded areas indicate 95% confidence intervals.

fore the interaction term between *industry diversity* and *WTO exposure.* Controls include the same province characteristics and the same time period as in the accountability analysis plus the province accountability measure. Figure 3.9 shows the positive interactive relationship between industry diversity and developmental responses across different levels of WTO exposure. Consistent with theoretical predictions, the figure indicates a positive relationship between industry diversity and the adoption of a developmental strategy in addition to a strengthening of this relationship at higher levels of WTO exposure.

While the rules and commitments governing China's accession to the WTO were designed to bind China more tightly to international rules grounded in liberal market-economy principles, this chapter reveals that the resulting policy changes have been far from uniform. Instead, the analysis highlights the importance of state structure in mediating responses to globalization. Rather than a unitary "China response" that one might expect given single-party rule, the evidence presented here emphasizes why the economic and bureaucratic impact of WTO entry is received differentially by actors sitting in different positions within China's hierarchical and decentralized governing structure, leading to a stark policy divergence.

The adoption of these divergent strategies, grounded as they are in distinct state-market relations, belies the inadequacy of framing state responses to international rules in dichotomous terms of compliance or defiance and of protectionism versus liberalization. As this chapter has underscored, actors within the state can—and do—choose from a range of different strategies when engaging with international integration. These strategic responses are driven by a combination of international and domestic factors. While deviating from the WTO's market-liberal principles generates a threat of sanction, the costs of deviation fall more heavily on the central government as sovereign representative compared to subnational authorities, giving the latter greater scope to enact nonregulatory strategies. The choice between developmental and directive strategies in turn depends on the diversity of a jurisdiction's industrial base. In general, larger and more diversified jurisdictions are better able to exploit the export opportunities from WTO entry by enacting developmental strategies, while poorly diversified units tend to provoke the deployment of directive strategies to fend off import competition.

Hence, we observe that although the central government intensified its market-enhancing regulatory strategy largely in line with the spirit of WTO rules, this regulatory response weakened at the provincial level and even declined in local administrations. Instead, provincial governments strengthened their deployment of the market-shaping developmental strategy to take advantage of export opportunities created by trade liberalization. Local governments focused on boosting the directive content in their industrial policies to more forcefully shape industry outcomes in their jurisdictions. Arguments about whether the WTO has succeeded or failed in shifting China toward more market-liberal forms of economic governance end up obscuring these internal dynamics. In contrast, a more disaggregated view of Chinese economic policymaking allows us to gain a deeper understanding of the conditions under which WTO entry does—and does not—trigger shifts toward greater market reform.

Chapter 4

Explaining the Rise of State Capitalism

Around 2009 the phrase *"guojin mintui"* became highly popular in Chinese discussions about the economy. Meaning "the state advances while the private sector retreats," the phrase captured rising concerns over the dominant role of the state and the increasingly powerful positions that state-owned enterprises (SOEs) were occupying in the economy. Concurrently, scholarly works in the West increasingly pointed to the rise of Chinese state capitalism, the strengthening of industrial policy, and a state-coordinated "China, Inc." form of governance, indicating a strengthening of market-shaping policies in the central government.[1] How does the turn toward developmental statism documented in these studies accord with the finding in chapter 3 on the central government's enactment of market-enhancing, regulatory policies after entry into the World Trade Organization (WTO)? If indeed there has been a rise in Chinese state capitalism, when did this trend begin to emerge, which part of the vast Chinese party-state was driving this shift, and what explains the timing of the shift in policy trajectory?

In this chapter, I show that the central government's regulatory response to WTO entry did not last. Instead, a shift toward developmentalism in the form of strengthened industrial policies emerged in the latter half of the 2000s.[2] As this chapter will explain, these shifting policy trajectories were driven by different sets of agencies within China's powerful central bureaucracy. In the initial WTO period, the broad liberalization of many economic sectors was supported by newly empowered regulatory agencies working to spur the expansion of market forces. This involved the creation of regulatory commissions, legal reforms to delimit the exercise of administrative powers, new laws on competition policy, and administrative reforms to strengthen vertical supervision structures.[3] A few years later, however, central policy came to be dominated by agencies such as the State-owned Assets and Supervision Administration Commission (SASAC), which governs central SOEs, the National Development and

Reform Commission (NDRC), and the Ministry of Industry and Information Technology (MIIT), which advocated a more market-shaping approach for the promotion of priority sectors (in particular high-tech industries).

This reversal in policy trajectory is the central puzzle of this chapter. First, why did the rise of regulatory institutions in response to WTO entry not constrain the belated strengthening of developmental policies? This phenomenon seems to call into question arguments in historical institutionalism about the lock-in and increasing returns dynamics that accompany the growth of new institutions.[4] Second, what explains the timing of this policy change? In the next section, I address deficiencies in the dominant explanations that have been put forth for the emergence of Chinese state capitalism in the latter half of the 2000s before laying out my own argument. The chapter then gives an overview of the shift in the central government's policy trajectory from market-enhancing liberalization to market-shaping developmentalism in the WTO period. Finally, I provide evidence for my explanation of the pattern and timing of this policy reversal, focusing on changes in the ability of the party to discipline the central state and how WTO rules altered the prospects for political advancement of different central agencies.

Dominant Explanations

There are two dominant explanations for the rise of state capitalism in China: the first points to the 2008 global financial crisis and the second to a change in elite preferences with the leadership transition of 2003. Each, I argue, is inadequate in addressing the WTO-era pattern of institutional change. The first explanation posits that the 2008 global financial crisis led the central government to apply a massive US$586 billion fiscal stimulus to the economy, much of which was channeled through SOEs and local governments.[5] However, this argument does not accord with evidence presented later in this chapter showing the rise of developmental strategies *prior* to 2008, particularly with the surge of industrial policies for high-tech innovation launched at the start of the Eleventh Five-Year Plan in 2006.

The second explanation posits that the 2003 change in leadership from Jiang Zemin and Zhu Rongji to Hu Jintao and Wen Jiabao fundamentally altered leadership preferences over economic governance. Proponents of this argument point out either that Hu and Wen did not believe in reform as strongly as Jiang and Zhu, that Wen was a strong supporter of state planning, or that Hu and Wen simply had different policy priorities such as expanding social insurance and poverty reduction.[6] This view is succinctly expressed by the Office of the United States Trade Representative (USTR), which releases annual reports on China's compliance with WTO rules. The 2012 report explained the rise in government intervention in the economy as follows:

In 2003, when new leaders took over in China, the Chinese government continued to take steps to implement the WTO commitments that China had agreed to phase in over time, furthering China's economic reforms. However, beyond these steps, China's new leaders for the most part did not continue down the path pursued by their predecessors. Beginning with the creation of the State-owned Assets Supervision and Administration Commission (SASAC) in 2003, China's new leaders de-emphasized their predecessor's move toward a greater reliance on market signals and instead set out to bolster the state sector.[7]

Scholars also point to the change in leadership as the critical turning point in China's shift toward a more activist industrial policy. In contrast to Jiang and Zhu, "the Hu-Wen administration [laid] a renewed emphasis on active state guidance and multi-year programs in economic, social and technological development. As a consequence, many previously influential protagonists and brokers of economic liberalization were sidelined, whereas indicative planners and industrial policy protagonists merged into a "centrist" or "statist" advocacy coalition that became the dominant force in economic policymaking.[8] Wen "put a renewed trust in the planners," and the bureaucracy was restructured accordingly to reflect these new political priorities.[9] Such views are broadly similar to the argument that different bureaucratic agencies serve as an informal power base for Chinese Communist Party (CCP) leaders. As such, the rise and fall of various agencies, along with their changing functions, reflects the shifting balance of power and preferences among conservatives and reformers at the top of the party structure.[10]

There are three main problems with such preferences-based explanations. First, by assuming that the pattern of policy change derives from leadership preferences, this logic deduces the cause of an outcome from the effect itself, implying that whatever type of change the leadership desires can and will happen. This framing ignores the role of the bureaucracy either in shaping the form and content of policies or in resisting the goals of the political leadership. By leaving out the channels through which bureaucratic agencies *resist* change, this framework cannot explain, if leaders can create and destroy agencies in accordance with their economic agenda, why some agencies are only downgraded after restructuring rather than completely abolished. For example, while Premier Zhu was able to abolish and subsume major industrial ministries under the State Economic and Trade Commission (SETC) in 1998 in his attempt to consolidate industrial policymaking, other industrial agencies remained untouched, such as the Ministry of Information Industry and the State Tobacco Monopoly Administration. Why were these left standing? Wen's 2008 bureaucratic restructuring established five "superministries" to consolidate policy making across a number of spheres.[11] Yet one proposal for a superministry governing energy was never implemented "due to strong opposition and lobbying from both the NDRC and energy firms."[12] In short, this preferences-based framework falls short in explaining when and why the political initiatives of leaders end up being circumscribed, partially successful, or failures from time to time.

Second, a preference-centric explanation places too much emphasis on differences in ideological leanings between the Jiang-Zhu and Hu-Wen administrations. This critique is not meant to dismiss the importance of elite politics or to suggest that individual leaders do not differ in their preferences either within or across administrations.[13] Rather, these differences do not have enough explanatory power to account for the pattern of policy change in China's WTO years. Instead of being highly statist, Wen held strong reformist credentials from his time as head of the Central Committee General Office for Hu Yaobang and Zhao Ziyang—two leaders who were both strong proponents of economic liberalization.[14] Wen was a crucial member of Jiang and Zhu's economic team and was placed in charge of the Central Financial Work Commission in 1998. Wen was also said to have had Zhu's support in replacing the latter as premier.[15] If the ideological differences between the two were so large, it is not clear why Zhu would have endorsed Wen and risked having his hard-fought reforms unraveled by his successor. In addition, the preferences-based explanation gives too much credit to Zhu as a driver of market liberalization and neglects other aspects of his economic agenda that were less market-friendly in nature (particularly in policies surrounding SOE management). The explanation similarly downplays the regulatory reforms that did take place under Wen's premiership.

Finally, by focusing only on the rise of state capitalism, explanations based on leadership preferences cannot explain the features of institutional change that emerged within the central state in China. Why did the rise of regulatory institutions in the years immediately after WTO entry not constrain the subsequent rise of developmentalism? Why wasn't the rise of state capitalism accompanied by the destruction or abolition of regulatory agencies? How did we end up instead with the strengthening of "rival" institutions at different points in time?

Institutional Change in an Authoritarian Regime

In addressing this puzzle, I build on the work of historical institutionalists in analyzing institutional change as a conflictual process.[16] While new policies can generate positive feedback effects whereby incentives and resources for mobilizing supporters lead to self-reinforcing institutional change, these same policies can also generate countermobilizations—losers from new policies do not simply disappear but can organize to regain lost ground.[17] I apply these analytical tools in examining how the introduction of WTO rules created different winners and losers within the bureaucracy, which then shaped the coalitions supporting different types of institutional consolidation in China.

However, an account that examines institutional change through the lens of feedback effects and coalitions alone cannot explain the pattern of change. What were the mechanisms that drove the shift in the central government's policy trajectory? How do the features of authoritarian policymaking interact with the international environment to affect the shape and timing of domestic institutional change?

Ability of the Party to Discipline the State

I argue that we need to situate institutional politics in the context of China's authoritarian regime and the power relations that structure its governance system. Shifts in power within the state have to be understood in the context of the bureaucracy's relationship with its political masters and, crucially, whether party leaders have the ability to discipline the state. One challenge in drawing such a clear distinction between the party and its state is that the dividing lines are in practice rather blurred in the Chinese context. However, while all top bureaucrats are party members (i.e., the state is a part of rather than separate from the party), there is still a division of responsibilities between the party leadership, which makes the most important political decisions, and the bureaucracy, which is charged with implementation. For analytical clarity, I define the party leadership as the members of the Politburo Standing Committee (PBSC) and the state as party members working in the central bureaucracy (ministries, commissions, etc.).

Political leaders commonly rely on the state bureaucracy to implement their goals. Dan Slater and Sofia Fenner have pointed out the important role that bureaucracies play in party-based authoritarian systems, arguing that "states are the ultimate institutional weapons in the authoritarian arsenal."[18] Political parties require a state apparatus to implement their strategies and goals, and therefore the durability of authoritarian regimes depends on the "infrastructural power" of the state.[19] I build on this concept by introducing greater specificity into the range of relationships that might exist between the state and the party. Slater and Fenner's formulation of state power focuses on the question of how *effectively* the state apparatus can carry out the goals of authoritarian regimes. This assumes that the state will always be equally motivated to carry out the goals of the party. Such automatic obeisance to party decisions should be questioned even in authoritarian regimes, since different political decisions (e.g., WTO entry) can generate winners and losers within the central bureaucracy, leading to both support and resistance. We therefore need to focus not just on state capacity but also on the political relationship between the party and the state. As Deng Xiaoping himself said, "Once a political line is established, someone must implement it. Depending on who is in charge of implementation—those who support the party's line, those who do not, or those who take the middle-of-the-road position—the results will be different.[20]

To capture the politics behind state-party relations, I build on Joo-Youn Jung's concept of "reciprocal reliance" wherein bureaucrats are "both the agents and the constituents of the top leaders."[21] In other words, while the bureaucracy is functionally responsible for implementing political decisions (and is therefore an *agent* of the party), the party leadership nevertheless has to rely on the bureaucracy's implicit support for these decisions to be assured that its orders will be carried out. Therefore, the party leadership also views the state as a critical *constituency* whose support it needs to actively court. This concept of a mutual give-and-take between the party and the state is important but again seems underspecified. While the two roles (being an agent versus a constituent) are arguably constantly in tension and always at play, when is the state on balance predominantly an agent of the party, and when does it largely play the role of constituent? How do policies change depending on where the balance of the relationship is?

I posit that one variable driving the balance of this relationship is the degree to which the support networks of the party leadership are *embedded* in the central bureaucracy. By embeddedness, I refer to the degree to which the career of a particular party leader was primarily spent in the central bureaucracy in Beijing versus in various prefectures and provinces across China. This embeddedness affects the likelihood that the party leadership has the autonomy to discipline the state when the latter deviates from its delegated tasks, such that the state acts as a faithful agent of the party, or whether the leadership is instead reliant on the bureaucracy because it is beholden to the state as a key political constituency.[22]

Leadership networks are an important component of governance in China, because unlike in democratic regimes or even semicompetitive authoritarian regimes, political competition does not take place between parties and is not structured by formal electoral rules. Instead, political competition takes place *within* the CCP, governed at best by nascent and underinstitutionalized rules. While one aspect of this contestation occurs, as I have noted earlier, through bureaucratic ranks in the cadre evaluation system, another aspect arguably takes place through factional politics especially at higher political levels.[23] Absent formal rules for political succession, elites compete for advancement via informal rival factions or networks. Contestation might be defined in terms of conservatives versus reformers, princelings versus Communist Youth League members, or generalists versus technocrats.[24] In this chapter, I focus on whether leadership networks are embedded within the central state or disembedded (i.e., rooted in the provinces outside of Beijing).[25] The politics of reciprocity permeates these networks, where the leaders of a faction rely on members to advance their political agenda while members rely on leaders for political advancement.

This dynamic leads to the following propositions: Leadership ties are *disembedded* from the central bureaucracy when the leadership's network of political support and mutual obligations built up over their career lies in the provinces and outside the central state. Under such circumstances the leadership, being in a position of relative autonomy, can effectively punish central bureaucratic actors who deviate from its policies without inflicting these punishments on members of its own network. As a result, the party leadership is able to effectively deploy the central state as its *agent* for implementing political goals.

In contrast, leadership ties are *embedded* within the central bureaucracy when leaders have risen up the political ranks largely through positions within the central state. Under such circumstances, leaders are unable to disentangle themselves from the interests of the bureaucracy or from the obligations and ties that have enabled their political rise. As a result, the party leadership, being in a sense captured by the central bureaucracy, cannot punish the state for deviating from its delegated tasks without also inflicting that same punishment on members of its own network. Instead, party leaders have to rely on the central state as a key *constituent* for shoring up their political strength.

The politics of this reciprocity generates outcomes that are distinct from what one might predict by simply examining the work experience of various Chinese leaders or by adopting a view of factions as operating in a top-down fashion. For example, a leader who has risen to the top through an extensive career in the central state might be expected to have in-depth knowledge of how the bureaucracy works and therefore be more adept at achieving his or her goals. Or, taking a top-down view of factions, one might expect a leader whose faction is based within the central bureaucracy to be more effective because his or her orders will be reliably carried out by faction members.

However, viewed through the lens of reciprocity, the factional relationship becomes more complicated. Obeisance from members of a network cannot be assumed, and the objectives of party leaders and the central state can diverge even within a faction. In cases where there is no conflict between the interests of party and state, we would expect a leader whose networks are embedded within the state to be highly effective in achieving his or her goals. In cases where a party leader wishes to adopt a policy that the bureaucracy resists, however, implementation requires the ability to discipline those within the state who deviate from their political instructions. The leader whose network is embedded within the bureaucracy becomes counterintuitively *less* able to achieve this goal, because any punitive action (e.g., investigation, demotion, or removal of authority over policy issues) would have the direct effect of weakening his or her own network.[26]

As such, depending on this degree of embeddedness, the party-state relationship that dominates (agent versus constituent) affects whether responses to WTO entry are driven by the party leadership or the central state. I argue that the Chinese leadership in the Jiang-Zhu era (1998–2003) was in a position of autonomy vis-à-vis the bureaucracy, as the political ties of Jiang and Zhu were largely *disembedded* from the central state. As such, even in the face of strong bureaucratic opposition, these leaders would be able to effectively discipline the state and be able to deploy the central bureaucracy as an agent to carry out their regulatory reforms. In contrast, the succeeding era of Hu and Wen (2003–2013) brought about a reversal in the political relationship between party and state.[27] Unlike Zhu, Wen's political networks were deeply *embedded* within the central state. Therefore, Wen would not be able to effectively punish the state for deviations from leadership policies and instead would need to rely on the state as a constituency to shore up his political strength. This distinction is consequential because, as discussed, the preferences of the party leadership across the two eras were not that different. The overall policy trajectory, I argue, depends on which agencies in the central bureaucracy were relatively more empowered, as the central government was split between agencies supporting regulatory versus developmental strategies.

WTO Leverage

The question of what types of strategic responses were advocated by which central agencies depends in turn on how WTO entry altered these agencies' prospects of political advancement. While chapter 3 assessed political advancement in terms of hierarchical rank promotion, advancement for central government actors refers to an expansion of agency influence over policy. Within China's authoritarian system, different central agencies are engaged in constant competition with each other for policy influence. I assume that each agency's approach to economic governance tends to be largely fixed over time such that particular sets of agencies tend to prefer to deploy a particular state strategy when advancing their policy agenda (see chapter 2 for an elaboration of this point). For example, technocratic agencies in charge of setting standards are likely to support regulatory strategies, while those charged with setting industrial policy are likely to be more developmental in orientation.

Whereas Theda Skocpol points out that new rules and policies have the potential to alter state capacities, I propose that new rules also have implications for the distribution of power within the state.[28] As explained in chapter 2, WTO entry affects prospects for bureaucratic advancement through both economic and bureaucratic channels. This chapter focuses on the bureaucratic channels through which WTO rules alter the ability of actors to maintain or enhance their policy influence. On the one hand, WTO rules threatened to circumscribe the discretion of some economic agencies, subject their policymaking process to greater public disclosure and external scrutiny, and transfer judicial authority over trade disputes to the WTO dispute settlement board. Such changes would pose concerns to any bureaucracy but to a far greater degree for an authoritarian bureaucracy used to operating in opacity and with weak formal constraints on its daily decision making. On the other hand, these same rules might be aligned with the interests of technocratic and legal agencies. The functional goals of these agencies are very much in line with the principles advocated at the WTO and hence their influence stands to be boosted by the

WTO's emphasis on international standards setting, intellectual property rights protection, and the rule of law. WTO entry might therefore provide these agencies with external leverage to promote their policy agenda, providing them with more opportunities for political advancement.

I propose that regulatory agencies in charge of standards setting are likely to see benefits from supporting WTO-led reform, as the implementation of technocratic WTO rules oriented toward building a regulatory state would enhance their policy relevance and influence. They are therefore likely to respond to WTO entry with stronger regulatory strategies. In contrast, agencies in charge of setting industrial policy are likely to be more developmental in orientation and hence would be opposed to WTO rules that threaten to circumscribe the set of policy tools at their disposal. They are therefore likely to seek opportunities to resist the imposition of WTO rules and look for opportunities to continue enacting developmental strategies.

The question of which set of agencies is in a stronger position to advance their agenda in turn depends on what I term "WTO leverage." I propose that in the early years of China's WTO entry, regulatory agencies whose authority was otherwise weak were able to use China's WTO commitments as leverage to push for greater regulatory reforms, improving their prospects for political advancement in the process. Importantly, the strength of this leverage weakened over time due to the detailed commitments documented in China's protocol of accession, which set specific dates by which various concessions and reforms would be completed.[29] As such, WTO leverage was highest at the point of WTO entry but gradually weakened over time as the implementation of China's WTO commitments progressed.

The institutional consequences of relying on external rules as a substitute for domestic support for reform was that there were only weak positive feedback effects supporting the strengthening of regulatory institutions as part of WTO entry. Consequently, those regulatory institutions ended up being supported only by a narrow coalition within the bureaucracy. Importantly, the "losers" from WTO entry did not disappear. Regulatory agencies were established *alongside* rather than replacing existing rival developmental agencies. Moreover, the strategy of relying on external rules to push domestic reform ended up generating *negative* feedback effects (i.e., active resistance to regulatory strategies) among these losers. As a result, a coalition emerged within the bureaucracy that became more anti-WTO and opposed to regulatory reform over time.

Therefore, as the strength of WTO leverage weakened, "losing" agencies supportive of developmental strategies would be able to consolidate their position and push for greater policy influence. As such, the implementation of WTO rules not only did not preclude but may also have generated conditions that led to the strengthening and consolidation of rival agencies advocating developmental approaches such as the NDRC. Consequently, developmental institutions saw a period of strengthening from 2006 onward after the implementation timetable in China's protocol of accession was largely completed.

Predicting Central Government Responses to WTO Entry

In sum, the embeddedness of party leadership networks affects the ability of the party to discipline the state and deploy it as a faithful agent and hence whether responses to the WTO are driven by the party leadership or the state bureaucracy. Concurrently, WTO leverage affects which central agencies were able to advance their policy influence at different points in time. Figure 4.1 shows how the two variables of party-state relations and WTO leverage combine to generate time-varying responses to WTO entry within the central government. I focus on regulatory and developmental strategies as the main potential responses that various agencies are likely to adopt, in line with the argument in chapter 3 that the central government is likely to respond to WTO entry with either regulatory or developmental strategies.[30]

The period immediately after WTO entry (2001–2003) is associated with high WTO leverage and a strong ability of the party leadership to discipline the state. As such, this period is likely to witness the strongest regulatory responses to WTO rules, driven by the leadership and bolstered by proreform agencies within the central government. This period is therefore likely to be associated with the active building of regulatory institutions.

From 2003 to 2005, the ability of the party to discipline the state weakened as Hu and Wen took over as leaders. However, WTO leverage was still relatively high during this period due to the schedule of implementation built into the protocol of accession. Therefore, while the party leadership would not be able to effectively drive liberalizing reforms, regulatory agencies within the cen-

Ability of party to discipline the state

		High Jiang-Zhu Era 1998–2003	Low Hu-Wen Era 2003–2013
Strengh of WTO leverage	**High** 2001–2005	Regulatory institution-building (2001–2003)	Dispute / Protocol-driven policy change (2003–2005)
	Low 2006–present	*Xi Jinping era?* *(2013–present)*	Developmental institution-building (2006–2013)

Figure 4.1. Predicted responses from the central government

tral government would still be able to use the protocol and WTO rules as leverage to advance their agenda. This period would therefore be associated not with institution building but instead with the continued promulgation of regulatory strategies driven by the protocol or by WTO disputes.

The period 2006–2013 is associated with a weak ability of the party to discipline the state and low WTO leverage as the implementation timetable in the protocol expired. As such, economic policy would likely have been driven by the state rather than the party and specifically by developmental agencies within the central government as regulatory agencies lost their policy leverage. Consequently, we would likely observe the strongest period of developmentalism in this period, resulting in activist developmental strategies and rival institution building.

The final potential scenario of low WTO leverage and a strong ability of the party to discipline the state falls outside the scope of this study, as the period of low WTO leverage (2006 onward) does not overlap with the Jiang-Zhu leadership (1998–2003). However, one might tentatively speculate that this period corresponds with the Xi Jinping era, since Xi has emerged as a powerful leader determined to discipline the party-state through a strong anticorruption campaign. I briefly discuss the implications of this framework for economic policy in the Xi era toward the end of this chapter.

I employ a mix of quantitative and qualitative methods to test these arguments. First, I track the changing intensity of state strategies deployed by different central agencies using the prevalence of different strategies found in their policy documents. This sheds light on which coalition of agencies is driving the expansion of central regulatory or developmental policies. To identify the turning point of when developmental strategies became ascendant, I examine the activism of various developmental agencies in issuing policies over time and assess the change in foreign firm sentiments on the business environment using data from an annual survey of US firms in China. I rely on qualitative analysis to identify changes in the political relationship between the CCP leadership and the central government. This involves examining the speeches and records of meetings by relevant leaders in the pre-WTO and WTO periods to assess the top leadership's approach to economic reform. These materials include their work reports at the National People's Congress, records (public and internal) of their meetings with Chinese and foreign officials and businesspeople, and internal memos sent to the leadership for their approval or comments. Other sources include newspaper reports (in English and Chinese), official CCP journals, and articles written by government officials in scholarly journals (e.g., the *Chinese Cadres Tribune*, the *Journal of the Central Committee Party School*, and the *Review of Economic Research*).[31] I also conducted interviews with current and former officials in the Chinese and US governments and the WTO as well as with industry association representatives with experience in engaging China's central economic agencies.

The Emergence of Institutional Dualism

As discussed in chapter 1, China's reform path has been marked by intense political and ideological contestation over the role of the state in governing the economy. Reforms reflecting new governance approaches were often carried out through newly established or reformulated agencies, while old approaches embodied in existing agencies did not necessarily go away and often remained in the bureaucracy, sometimes in diminished form. The experimental and incremental nature of China's reform process therefore meant a bureaucratic landscape that was similarly varied.

The locus of authority over industrial policy has also shifted within the bureaucracy over the post-Mao reform era, mainly between two competing agencies: the State Planning Commission (SPC) and the State Economic Commission (SEC). The rivalry between these two agencies (and their later incarnations) has persisted over time even as the central bureaucracy has gone through various restructurings under different leaders. Functionally, each agency had slightly different responsibilities: the SPC held authority over economic planning and project approval, while the SEC oversaw SOE governance. This division inevitably generated competitive tensions, because questions of economic growth and planning could not be separated from issues of SOE governance.[32] Each agency also tended to push for different approaches to development. As the SEC was responsible for reforms introducing greater enterprise autonomy, it came to be seen as a leading reform agency, while the SPC was perceived more as a conservative agency attached to the planned economy mind-set.[33]

In 1988, a round of administrative restructuring launched by Premier Li Peng led to the abolition of the SEC and the absorption of its functions into the SPC. When Zhu was put in charge of economic reform in 1991, he took over an outfit called the Production Office, which was upgraded to the Economic and Trade Office in 1992 and further elevated to the SETC in 1993, in essence a reincarnation of the SEC.[34] The SETC was put in charge of industrial policy and SOE reform, advocating a developmental approach that was seen to be more reformist and market-friendly compared to the SPC.

When Zhu rose to the premiership in 1998, he immediately set about another round of bureaucratic restructuring, this time to substantially trim the size of the central bureaucracy to make it more rational and efficient. The number of ministries and commissions was cut from forty to twenty-nine, and personnel was reduced from 8 million to 4 million.[35] Over 230 officials of vice ministerial rank were laid off.[36] In this downsizing, ten different industrial agencies were abolished and subsumed under the SETC.[37] The SPC was downgraded to a research institute and its name changed to the State Development and Planning Commission (SDPC). The newly consolidated SETC gained the status of a superministry in charge of industrial development and was informally referred to as the "mini–State Council."

The Rise of the Regulatory State

In the pre-WTO period, agencies in charge of regulatory standards were weak and below ministerial rank and did not play a significant role in economic governance. In the second half of the 1990s under Zhu's watch, reforms were instituted to strengthen the regulatory functions of the central government.[38]

While WTO entry was no doubt a strong driver of these regulatory reforms, it was by no means the only impetus. Domestic pressures for stronger regulatory standards were also building due to scandals involving workplace and consumer safety.[39] In 1999, the central government made changes to strengthen its vertical control over the State Administration of Quality and Technical Supervision. This agency was then upgraded to ministerial status and renamed the State Administration of Quality Supervision, Inspection and Quarantine (SAQSIQ), consolidating within its ambit responsibilities for standards supervision, inspection and quarantine, certification, accreditation, and quality control as well as regulatory functions that had previously been spread out across industrial agencies.[40] These broadened responsibilities made the SAQSIQ "one of the main arms of the regulatory state."[41] Figures 4.2 and 4.3 shows the clear increase in the number of policies and measures issued by the SAQSIQ as well as an intensification of regulatory language in those documents in the WTO period, evidence pointing to the rise of a stronger, more centralized regulatory state.

The growing centralization of regulatory functions took place as well in other agencies but to a weaker extent.[42] These included the State Administration for Industry and Commerce (SAIC), in charge of business registration, trademarks infringements, business fairs, and protecting consumer interests; the State Environmental Protection Administration, now the Ministry of Ecology and Environment; and the State Drug Administration, later renamed the State Food and Drug Administration. Market-enhancing reforms also involved an intensification of legislative activity to make China's legal system compliant and consistent with WTO articles. Outside of sector-specific regulations, sweeping changes were made to the Foreign Trade Law, the Customs Law, the Product Quality Law, the Pharmaceutical Administration Law, the Copyright, Patent and Trademarks laws, and more.[43] Furthermore, the domestic administrative and judicial systems were substantially restructured to allow for greater transparency, channels for judicial review, and more uniform administration.[44]

These changes added up to a substantially transformed system for economic governance that was not only more liberalized in terms of tariffs and trading rights but also more transparent, undergirded by a set of WTO-consistent laws and, most importantly, governed by a set of newly empowered regulatory institutions. Given the momentum behind these market-enhancing policies, how and why did a resurgent "state capitalism" emerge a few years later?

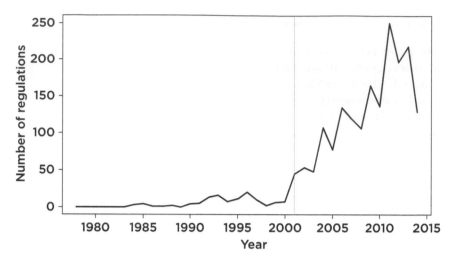

Figure 4.2. Number of sector-specific regulations issued per year by the SAQSIQ
Source: Author's data set.

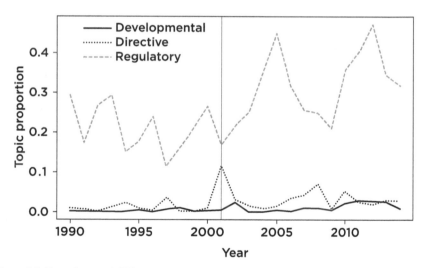

Figure 4.3. Prevalence of different strategies in SAQSIQ regulations (including the China Inspection and Quarantine bureau and the Certification and Accreditation Administration)
Source: Author's data set.

The Rise of the Developmental State

One interviewee, an official in an international organization, observed that the turning point in China's policy trajectory seemed to occur around 2006 and 2007, with a rise in negative sentiments among China's trading partners in the WTO. Not only had there been no further liberalization in policies in recent years, but SOEs had become more dominant in the economy. SOE reforms had been limited to improvements in corporate governance, with little movement toward privatization or liberalization.[45] As another sign of potential discontent against Chinese industrial policy, the number of complaints filed against China at the WTO's dispute settlement board rose sharply in 2006 and has not abated since. While China has also become more active as a complainant in the dispute settlement system, it continues to be a more frequent respondent to complaints filed by other members.[46]

This rise in WTO complaints accords with a negative turn in the sentiments of foreign businesses in China around the same time. Figure 4.4 shows how US firms in China have perceived the impact of Chinese economic reforms on their business climate. The chart shows that US firm sentiments were highly positive between 2002 and 2004, with those saying that reforms had benefited the climate for US businesses "to a very great extent" or "to a great extent" ranging between 61 percent and 69 percent. From 2005 onward there was a sharp drop in those positive sentiments, with the same responses ranging between 30 percent and 44 percent. In 2006 sentiments were the most negative, with only 7 percent of respondents saying that US firms had benefited "to a very great extent" from recent reforms.

Moving in step with these increasingly negative sentiments, the USTR's annual reports to Congress on China's WTO compliance became sharply more critical from 2006 onward, with the 2006 report quoting one trade association official as saying that "over the past 12 months we have seen an upsurge in industrial planning measures as tools of economic development by central government authorities."[47] The 2007 report also quoted critical remarks from a trade official, noting that "some policy makers also appear to want to expand the [Chinese] government's role in directing the economy and in developing internationally competitive Chinese enterprises, while also restricting the role of international companies in certain sectors. Designation of 'pillar' industries, promoting 'indigenous innovation,' and establishing 'national economic security' criteria to review deals are troublesome signposts that do not imply full market access for U.S. companies."[48]

The policy developments highlighted in the 2007 USTR report came from China's Eleventh Five-Year Plan and associated policies launched in 2006, which established "indigenous innovation" as a key development priority for the country with particular focus on industries such as new-energy vehicles, high-end equipment manufacturing, biotechnology, and information technology.[49] In 2005 and 2006, SASAC chairman Li Rongrong listed seven strategic

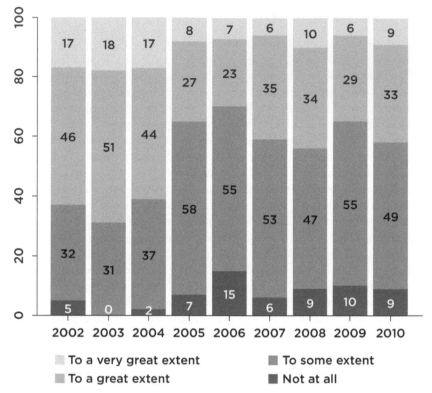

Figure 4.4. Change in business climate sentiments of US firms in China
Source: American Chamber of Commerce Business Climate Surveys, 2002–2010.
Note: Respondents were asked "To what extent do you believe China's recent economic reforms
have improved the climate for US business in China?" Numbers in bars indicate percentage of
respondents in each category. The "Not at all" category in 2006 included those who answered
"Don't know/No basis to judge."

sectors that comprise "the vital arteries of the national economy" over which
the state needed to maintain "absolute control."[50] Additional sectors were
designated "pillar" industries deserving of particular support in which important "backbone" enterprises would maintain strong controlling power.[51] The
emphasis on indigenous innovation marked a turn in industrial policy toward
what some have called "techno-nationalism," reflecting a desire by the central
government to deploy more market-shaping measures to push the Chinese
economy up the value chain beyond its traditional strengths in manufacturing, assembly, and other types of processing trade.[52] This new policy focus can
be clearly seen in figure 4.5, which shows the prevalence of the word "innovation" in industry regulations issued by the central, provincial, and local governments. The central government's use of the word in its regulations first

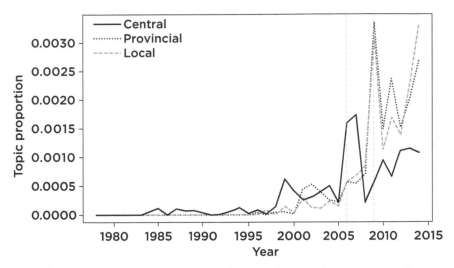

Figure 4.5. Occurrence of the word "innovation" in central, provincial, and local regulations
Source: Author's data set.

spikes in 2006 during the launch of the Eleventh Five-Year Plan and again in 2007 when additional policies associated with the Five-Year Plan were rolled out. As the effects of these policies reverberated to lower administrative echelons, provincial and local governments followed suit with dramatic increases in use of the word "innovation" in their policies from 2009 onward.

The emphasis on this innovation being indigenous was accompanied by policies that appeared to discriminate in favor of domestic enterprises. In 2006, the State Council released its Medium to Long-Term Plan for the Development of Science and Technology (2006–2020), which established reducing the country's dependence on foreign technology as a national goal. Eleven different industries were identified for priority development. As an example of the discriminatory aspects of this push for indigenization, the Ministry of Finance issued a number of policies in 2007 that required governments to give priority to "indigenous innovation" products when making procurements.[53] In 2009 additional policy focus was placed on developing "strategic emerging industries," heightening state activism in technology innovation and in industries such as new-energy vehicles, high-end equipment manufacturing, biotechnology, and information technology.[54]

Two ministries have come to be associated with the rise of these techno-nationalist policies in the WTO era: the NDRC and the MIIT.[55] Each was formed through a different round of administrative restructuring. When Wen took over the premiership in 2003 and reorganized the central bureaucracy, the SETC's responsibilities over industrial policy shifted back to its rival the SDPC, which also absorbed the functions of the State Council Office for Economic Restructuring and then was renamed the NDRC. The SETC's functions for SOE management was carved out into the newly created State-owned Assets Supervision and Administration Commission (SASAC), while SETC responsibilities over trade were absorbed into the Ministry of Foreign Trade and Economic Cooperation, which was then renamed the Ministry of Commerce (MOFCOM).[56]

The creation of SASAC in 2003 was an important milestone in SOE governance in China. By the end of the 1990s, SOE asset management had been recentralized to Beijing as part of the campaign to reform and streamline the SOE sector. However, governance responsibilities remained fragmented across five different central commissions and ministries.[57] The establishment of SASAC therefore represented a major consolidation of authority within Beijing over China's SOEs. One of the organization's objectives was to concentrate state assets into large holding companies by merging or restructuring smaller, less profitable entities. As part of this process, SASAC has gone from overseeing 196 central SOEs in 2003 to 96 in 2019.[58] While SASAC has arguably grown into a very powerful and consequential body, it is not closely involved in issuing industry-specific regulations across the economy. Instead, SASAC's influence is seen to be stronger in defending the party's ownership of assets in what are considered key strategic or "lifeline" sectors of the economy.[59]

Therefore, the clearest winner from this round of restructuring was the NDRC (formerly the SDPC and SPC), with newly consolidated powers over industry governance and with its old rival the SETC abolished.[60] One important source of the NDRC's strengthened influence comes from the authority that it gained over investment approval.[61] In 2008 another round of administrative restructuring led to the creation of the MIIT, formed by merging several different agencies.[62] The MIIT then became a superministry in charge of information technology development. Given the administrative complexity of these changes, figure B.1 in appendix B provides a summary timeline of the major changes brought to these various economic agencies as a result of different rounds of administrative restructurings.

My interviews with scholars, government officials (active and retired), and industry representatives affirmed the competition between these two approaches to economic governance within the central bureaucracy. There was broad consensus that the NDRC was the most conservative ministry when it came to reform, with interviewees referring to how the agency's roots in the SPC meant that NDRC officials still tended to rely on administrative guidance in their work, issuing orders from on high and attempting to mold an industry into what they wanted to see.[63] The MIIT was similarly described as a developmental agency that while not rooted in a planning mind-set was nonetheless highly nationalistic and looked to the Japanese experience of growth to try to leapfrog into more advanced technologies and industries.[64] The most liberal ministry was MOFCOM, whose staff tended to support greater marketization and openness in the economy and for China to keep on implementing reforms and to eventually join the Trans-Pacific Partnership (TPP).[65] However, interviewees also noted that MOFCOM was weak compared to the NDRC and the MIIT, as it had weak approval authority and little control over the allocation of resources, unlike the NDRC's strong control over investment approval.[66]

Explaining the Rise of State Capitalism

Two important trends mark the evolution of China's economic institutional and policy landscape in the WTO years. The first is a growing centralization of economic control, with tightened vertical control and elevated administrative status given to regulatory agencies such as the SAQSIQ in the immediate WTO years. The second is the consolidated control and greater activism in industrial policy by developmental agencies such as the NDRC and the MIIT from 2006 onward. To understand the timing and emergence of this state capitalism, we need to consider important shifts in party-state relations and WTO leverage with attention to the positive and negative feedback effects that drive institutional change. The degree to which political leadership ties are embedded within the central bureaucracy determines the likelihood that the party leadership can deploy the central state as a faithful agent, as opposed to having to rely on it as a key political constituency. The resulting party-state relationship that prevails (state as agent versus constituent) affects whether policy change is driven by the party or the state. At the same time, the relative empowerment of regulatory versus developmental agencies to influence the policy agenda depends on the strength of the leverage to be gained from WTO rules.

Jiang-Zhu Era: State as Agent under Strong WTO Leverage

In the Jiang and Zhu period (1998–2003), political leadership ties were relatively disembedded from the central state, allowing Jiang and Zhu to deploy the bureaucracy as an agent to achieve their goals of using WTO entry to drive domestic economic reform forward.[67] The degree of embeddedness of Jiang and Zhu's leadership ties can be assessed by examining each politician's career trajectory up to the point of his appointment to the party leadership (the PBSC).[68]

Jiang joined the Communist Party in 1946 and after the communists' victory over the nationalists in 1949 spent six years in Shanghai in various factory positions and the Shanghai division of the First Machine-Building Industry Ministry. He then trained for a year in 1955 at an automobile factory in Moscow before spending six years in Jilin province in northeastern China in various positions at the First Automobile Works plant. From 1962 to 1980 he worked in the First Machine-Building Industry Ministry in Shanghai and Wuhan and as deputy director and director of the ministry's Foreign Affairs bureau. It is unclear how many of his years in this ministry were spent in local bureaus outside of Beijing. I assess that at minimum he started out in Shanghai in 1962 and would not have been posted to the central ministry until after the Cultural Revolution ended in 1976 and those who had been "sent down" to the countryside could return to urban areas. In 1980, Jiang was appointed to the central government's State Administration Commission on Import and Export and the State Administration Commission on Foreign Investment. From 1982 to 1985, he held positions in the Ministry of Electronics Industry as deputy party secretary and deputy minister and subsequently minister and party secretary. He served as Shanghai's mayor and party secretary for the next four years and returned to the capital in 1989 when he was elevated to the PBSC. Altogether, Jiang's career prior to being appointed to the PBSC involved approximately thirty-two years outside the central bureaucracy and approximately eight years in it.

Zhu joined the Communist Party in 1949 and after graduation from Tsinghua University spent a year in northeast China in the planning bureau before moving back to Beijing from 1952 to 1958 to work in the State Planning Commission. In 1958 he was targeted in the Anti-Rightist Campaign for criticizing Mao Zedong's economic policies and exiled to the remote northwest regions of China to teach at a party school and then to work as an engineer in the Economic Integration Bureau for twelve years.[69] From 1970 to 1975 (during the Cultural Revolution) Zhu was sent down to the countryside to work in a cadre school. It was not until 1975 that he returned to the central government to work in the Ministry of Petroleum Industry for four years. He was moved to the SEC in 1979 and rose through the ranks over the next eight years, ending up as deputy party secretary of the SEC before moving to Shanghai to serve as mayor and party secretary from 1987 to 1991. He was appointed State Council vice premier from 1991 to 1992 and elevated to the PBSC in 1992. Altogether, Zhu spent twenty-two years outside the central bureaucracy and nineteen years in it before being appointed to the PBSC.

One consequence of Jiang's and Zhu's career trajectories is that neither of them had strong support networks within the central bureaucracy. Jiang's network, built up over the years, drew from his career outside Beijing and particularly in Shanghai.[70] Zhu, possibly as a result of his seventeen years in the political wilderness during the Anti-Rightist Campaign and the Cultural Revolution that followed, combined with his abrasive working style, never built up a strong support base or following within the party.[71] From one perspective, this lack of central government networks might seem a political disadvantage. However, the ties of mutual obligations built into a network relationship means that having strong ties embedded within the central bureaucracy can turn into a handicap, as it is then difficult to discipline the bureaucracy without also imposing those punishments on members of your own network.

While the observable implications of a state-as-agent relationship are difficult to measure, the strength of the Jiang-Zhu leadership can be seen not by their ability to embark on centralization initiatives that bureaucrats in Beijing would have benefited from, such as the 1994 fiscal reforms that rejuvenated central revenues, but instead from their ability to undertake reforms that directly *undermined* the interests of the bureaucrats. The first instance of this was in Zhu's 1998 administrative restructuring, where about half of the civil service was laid off and the number of ministries was slashed from forty to twenty-nine. A second instance was the SOE reform launched in late 1995 (when Zhu was vice premier) to consolidate and corporatize large SOEs and relinquish control over small and medium SOEs. This led to a dramatic fall in the number of SOEs, from 127,600 in 1996 to 34,280 by 2003.[72] The third instance was the push for China to join the WTO, which threatened to severely limit the administrative discretion of powerful central agencies.

Strong WTO Leverage, Weak Positive Feedback Effects

In the period running up to WTO accession, Zhu's domestic support within the central government for liberalization was weak.[73] However, as has been documented by other scholars and widely confirmed across my interviews, Zhu used the negotiation process and WTO commitments as external leverage to push domestic reform forward.[74] In addition, the necessity of negotiating entry into a major international organization strengthened the otherwise weak position of those within the bureaucracy who supported Zhu's efforts (i.e., agencies such as the trade ministry and SAQSIQ that benefited from WTO-led liberalization).

The timetable specified in China's protocol of accession for implementing its numerous WTO commitments provided additional leverage for proliberalization agencies in the years immediately following WTO entry. One interviewee noted that this external leverage was stronger in China than in the United States because in China international law was stronger than domestic law, while the opposite was true in the United States.[75] Many commitments, such as tariff reductions, services liberalization, and the liberalization of distribution rights, were phased in over a number of years, commonly to be completed by 2004 or 2005. For example, Annex 3 of the protocol lists the liberalization schedule for products subject to import licenses, quotas, and tenders. These nontariff measures were to be eliminated at specific intervals: immediately upon accession for granulated sugar, by 2002 for sodium cyanide, and by 2004 for aviation kerosene. Therefore, over time this external leverage weakened as the timetable for implementation reached its conclusion. As another interviewee pointed out, with the commitments in the WTO accession protocol largely met by 2005 and 2006, it was difficult to mobilize the bureaucracy to continue with additional reforms.[76]

In other words, the expansion of regulatory institutions was achieved through a continual reliance on WTO rules as external leverage, with international rules serving as a *substitute* for domestic bureaucratic support. As a result, the positive feedback effects that are posited to reinforce institution building were at best weak. One source of positive feedback to support institution building is the channeling of incentives or resources to certain groups, which has the reinforcing effect of creating new supporters of the policy reform.[77] However, the (somewhat ironic) feature of market-enhancing reforms to build regulatory institutions in China is that these reforms were designed to *remove* the bureaucracy's control over resource allocation. Supporters of regulatory reform in China, such as the SAQSIQ and MOFCOM, were therefore weak—and remained weak—precisely because they did not control substantial resources.

Another source of positive feedback to support institution-building might be the transformation of state capacities or skills.[78] While the pool of technocratic knowledge and skills surrounding international trade institutions certainly expanded in the initial WTO period, the overall base of knowledge within the bureaucracy remains narrow and has arguably even declined over time. Part of this diminishing knowledge is linked to the stagnation of the trade reform agenda at the WTO itself. The Doha Development Round, launched in 2001, failed to reach agreement in 2006 and continued to be met by failure in subsequent WTO meetings afterward.[79] As a result, there was no new round of external commitments that agencies such as MOFCOM could use as renewed leverage to continue advancing their policy agenda. The lack of progress in WTO rule making had its impact on the Chinese bureaucracy. One interviewee, a former government official, noted the diminishing regard for WTO-related issues in cadre training such that new generations of local officials had neither knowledge of nor appreciation for WTO rules. The interviewee pointed out that in part this was due to the marginalization of the WTO at the international level, which has led cadres to be more attentive to trade agreements such as the TPP and the Regional Comprehensive Economic Partnership.[80] In contrast, the period running up to and immediately after WTO accession was accompanied by a minor boom in WTO-awareness training sessions and WTO-related academic events and publications. WTO centers also were set up in Shanghai, Beijing, and Shenzhen to provide technical support and services related to WTO legal and trade matters.[81] The current state of affairs, however, was such that agencies such as the MIIT and the NDRC took neither WTO rules nor the need to comply with international obligations into account when setting industrial policies and left MOFCOM to deal with any problems that might later emerge.[82]

My interviews further highlighted the continued reliance on external rules as leverage for reform in the context of otherwise weak support for regulatory liberalization. One interviewee noted that in the context of poor awareness and appreciation within the bureaucracy for WTO rules, dispute settlement cases were in a sense "good" for China, as they prevented further marginalization of the WTO agenda within the government. Only when a complaint is filed at the WTO is there recognition of these international rules, and only then can change be pushed via the implementing requirements after a panel ruling.[83] Another interviewee explained that dispute and antidumping cases provided an opportunity for MOFCOM officials to engage in positive persuasion of other agencies on the importance of WTO rules and help spread the reform mind-set within the bureaucracy. For example, when a case is brought against China, this allows MOFCOM officials to engage with their counterparts in the industrial bureau of the MIIT or the NDRC and make the case that WTO rules provide a level playing field for China to engage with its trading partners. These cases allow MOFCOM officials to stress to other agency officials that the WTO process is objective and that those rules can in fact work in China's interests because China can use the same measures against other countries and in so doing protect its own interests.[84]

Hence, one important consequence of relying on external rules as a substitute for domestic support is that strategies to strengthen regulatory governance in China ended up being weakly supported domestically by a narrow coalition of actors such as the SAQSIQ and MOFCOM. Because the main impetus for regulatory reform was external, the positive feedback effects that would have supported broader institutional reproduction within the central bureaucracy ended up being relatively weak. As a result, support for regulatory governance not only did not spread, but other parts of the bureaucracy ended up becoming more opposed to rather than supportive of regulatory reform, as the next section will show.

Losers and Growing Resentments Generated from the WTO Process
One important dimension that fed into the rise of state capitalism is that the losers in each round of administrative restructuring did not go away but instead survived and could rise again as circumstances changed.[85] There were many such losers from Zhu's 1998 round of government restructuring, chief among which were the industrial ministries that were subsumed under the SETC and the formerly influential SPC, which was renamed the SDPC and "relegated to being a research institute."[86] Not only did the SDPC persist (albeit in diminished form), but the developmentalist school of thought was also actively being promoted and passed on to new generations of cadre members in economics curricula at places such as Renmin University.[87]

WTO entry further enlarged the group of losers within the bureaucracy. The substantial legal reforms that accompanied China's WTO accession placed a new layer of legal constraints over what the bureaucracy could and could not do. In particular, the process of WTO implementation generated negative feedback effects that undermined the reform process by instilling grievances among important ministries in the central bureaucracy. As one interviewee recounted, Zhu's approach of keeping negotiations restricted to a small set of people necessarily made agencies feel left out of the process. Yet when it came to having to implement WTO commitments, these same agencies were suddenly confronted with a set of substantial demands to execute. The fact that MOF-COM was in charge of the WTO implementation process further generated resentment among bureaucrats in those agencies, who perceived themselves as having to be subservient to MOFCOM and to work *for* MOFCOM.[88]

An additional aspect of WTO entry that generated negative feedback effects and built up resentment among the losers of reform was the transitional review mechanism that allowed WTO members to regularly assess China's WTO implementation process over a period of ten years. This review mechanism was set up especially for China and was entirely distinct from the WTO's occasional trade policy reviews. One interviewee pointed out that each transitional review mechanism process generated many questions from China's trading partners that the government then had to address. Hence, agencies such as the Ministry of Agriculture, the SDPC/NDRC, and the Ministry of Finance, used to operating domestically with absolute autonomy, suddenly found themselves having to account for their actions to an external audience. They were naturally unhappy with such "interference" and came to see the WTO implementation process as being opposed to their work, giving them more reasons to resist reform. The fact that the transitional review mechanism was created only for China, combined with the other WTO-plus commitments contained in China's protocol of accession, generated grievances within the bureaucracy, as China's "payment for entry" into the club was so much higher than for other countries.[89] The interviewee pointed out that China's bound tariffs were already much lower than those of most developing countries such as India and Brazil, and furthermore, China's applied tariffs were very close to its bound tariffs, unlike other developing countries. In short, the leadership-driven nature of China's WTO accession combined with the large adjustment burden that the bureaucracy then had to undertake led to a backlash within the central government. Some even said that Long Yongtu (China's chief negotiator for WTO accession) was a traitor to the nation.[90]

This hostility within certain quarters of the bureaucracy to international trade rules and external scrutiny was further underscored by an incident recounted by an interviewee. He described how, when on an investigation visit to a coastal province in 2008 (seven years after WTO entry), the provincial head of the NDRC pounded his fist on the meeting table and said that MOFCOM was "selling out" the country and that what he (the NDRC) said was what counted, that he (the NDRC) was in charge of the industry—and not anyone else.[91]

The process of joining the WTO and implementing China's commitments was therefore one balanced on a very narrow set of bureaucratic constituents, pushed forward by leveraging on external rules as a substitute for domestic support. However, this very process generated the seeds for opposition in the future. The bureaucratic impetus to push back against reform only increased in the WTO era, given that the agencies that were the losers from the double whammy of the 1998 government restructuring and 2001 WTO entry did not disappear but in fact gained more reasons to mobilize through a series of negative feedback effects generated by having to adjust to China's WTO commitments. The narrow base of support on which regulatory reform was founded allowed strategic space for the anti-WTO coalition to build up rival institutions at a later point in time.[92]

Hu-Wen Era: State as Constituency under Weak WTO Leverage

The most important change that the transfer of leadership authority from Jiang and Zhu to Hu and Wen in 2003 brought about was not a change in preferences at the top but rather the shift in the political relationship between party and state. In contrast to Zhu, Wen started his career with fifteen years in Gansu province doing geological and survey work, slowly rising in the provincial Geological Bureau before being posted to Beijing in 1982. From then on he spent a continuous twenty years rising through the central bureaucracy, moving from the Ministry of Land Resources to the politically important and sensitive Central Committee General Office and then on to the Party Secretariat and Politburo before being elevated to the PBSC in 2002. Due to his fairly low rank during his time in Gansu, Wen may not have developed very strong ties in that province, in contrast to other leaders who spent time in provinces as either governors or party secretaries. Notably, Wen was the only PBSC member during his time to have had no experience with heading a province.[93] As a result, most of his network of support would have been deeply embedded within the central state.

In contrast to Jiang, who spent approximately thirty-two years of his career prior to the PBSC in positions outside the central government, Hu spent between seventeen and nineteen years in the western provinces of Gansu, Guizhou, and Tibet. He began his CCP career in 1964 as a political instructor in Tsinghua University in Beijing and from 1968 spent thirteen years in Gansu province, first in a local bureau of the hydroelectric power ministry followed by various positions in the Gansu provincial government. He was appointed in 1982 to leadership positions in the Communist Youth League, in 1985 to the position of party secretary of Guizhou province, and then as party secretary of Tibet in 1988. According to reports, he spent only a year and a half of that four-year posting physically in Tibet and the rest of the time in Beijing.[94] In 1992, he was appointed to the PBSC. Altogether, Hu spent far fewer years outside of the capital prior to joining the top leadership and worked in relatively poorer and less politically influential western provinces (in contrast to Jiang's long career in the economic powerhouse of Shanghai). Hu's political networks are therefore likely to be more evenly divided between the central bureaucracy and the provinces. However, he clearly rose to the general secretary position with far weaker networks outside the central bureaucracy compared to his predecessor Jiang.

Other scholars have pointed to features of the leadership transfer from Jiang to Hu as revealing of the latter's political weakness. For example, Hu took over only two out of three of the top leadership positions in the transfer of power (general secretary and president), while Jiang held on to the party chairmanship of the Central Military Commission until 2004. The PBSC was further expanded from seven to nine members, with the two additional members said to be strongly allied with Jiang. Finally, the Sixteenth Party Congress marked a victory for Jiang rather than for Hu through its endorsement of Jiang's "Three Represents" in the party charter.[95] More recently, scholars have pointed out that while Jiang and Xi (Hu's successor) were both given the designation as being the "core" of the party's collective leadership, that term was never associated with Hu. That Hu never attained core status has been taken to reflect his political weakness.[96]

In addition to these political vulnerabilities, the new Hu-Wen leadership team was further hobbled by its deep networks embedded within the central government. These ties meant that the leadership could not discipline the bureaucracy without simultaneously hurting members of their own network. Instead, they had to rely on the state as a key constituency for their political strength. While this concept of embeddedness is difficult to measure, one way to compare this aspect of the two leadership teams is to assess the degree to which each was able to carry out their administrative restructuring plans. On this score, Wen's 2003 restructuring was far less successful than that of Zhu. At the start of Wen's premiership, expectations were high that due to his deep experience in the central government, he would be off to a "fast start."[97] In early 2003, media reports on Wen's restructuring suggested that the number of ministries would be cut from twenty-nine to twenty-one or twenty-two. However, the restructuring that finally emerged only reduced the number of agencies from twenty-nine to twenty-eight, suggesting that Wen was unable to overcome bureaucratic resistance to his goals.[98]

Hence, while both the Jiang-Zhu and Hu-Wen periods have been described as periods of centralization, the two periods were in fact different *types* of centralization. Under Jiang and Zhu, the party *disciplined* the state and pushed reforms to make the central state more efficient and streamlined. Under Hu and Wen, the party *relied on* the state, leading the central government to be able to strengthen its promotion of industrial policies. One interviewee affirmed this view, pointing out that industrial policies were more dominant after 2003 because Wen was weaker and therefore needed to work *through* central agencies.[99] Wen's reliance on the government bureaucracy is reflected in his "consensual approach" to policymaking, which some have criticized as resulting in "diffuse and uncoordinated" governance.[100]

The shape and content of industrial policy in the Hu-Wen era therefore tended to be driven by the bureaucracy rather than by the political leadership. The backlash within the bureaucracy from having to implement China's WTO commitments meant that these losers were primed to advocate for and support alternative governing institutions when the opportunity arose. With the shift in party-state relations, the losers from WTO entry—faced with a low likelihood of sanction from party leaders and little opposition from regulatory agencies due to the weakening of WTO leverage—were able to consolidate their position and engage in rival institution building, advocating for a more developmental approach to governing the economy.

Regulatory agencies were unable to push back against this due to the narrow base of bureaucratic support for liberalizing reforms and the gradual decline in WTO leverage over time. One interviewee noted that in order to facilitate WTO negotiations, a coordinating body headed by a vice premier (who ranks above ministers) was established to overcome opposition from various quarters in the bureaucracy. This coordination body was abolished after WTO accession, leaving proliberalization agencies such as MOFCOM without the higher-level authority needed to push reform forward. The interviewee noted that unlike the USTR's office in the United States, which has a direct reporting relationship to the president, MOFCOM was weak within the Chinese bureaucracy and did not have as much influence as the NDRC.[101]

The heightened bureaucratic activism in industrial policies during the Hu-Wen era can be observed in figure 4.6, showing the number of regulations issued per year in the main central agencies in charge of industrial policy: the SETC, the NDRC, and the MIIT. The number of regulations issued each year by the State Council is also included as a benchmark. From the figure, we can observe that while the number of State Council regulations being issued stays fairly stable over the years, there is a surge in NDRC policies after 2003, when the administrative restructuring abolished the SETC and responsibilities over industrial policies shifted back to the NDRC. There is another even larger surge in policy activity after another round of restructuring in 2008 when the MIIT was formed to take charge of industrial policy in developing China's information technology capabilities. These trends provide evidence in support of the argument that developmental agencies took on a much more activist stance in the Hu-Wen era.

Could the rise of state capitalism during Hu and Wen's tenure be explained by their focus on developing the western provinces and regional redistribution? While the Hu-Wen leadership rolled out a number of policies aimed at reducing the inequalities between China's urban coast and its rural interior, the dynamics shown in this chapter in fact demonstrate that the bureaucracy did not lean toward this leadership priority. Rather than focusing on regional redistribution or industrial policies suited to the development of Western provinces, developmental agencies in Beijing doubled down on efforts to promote strategic emerging industries and indigenous innovation—initiatives biased toward already more advanced coastal provinces that would exacerbate rather than mitigate regional inequalities.

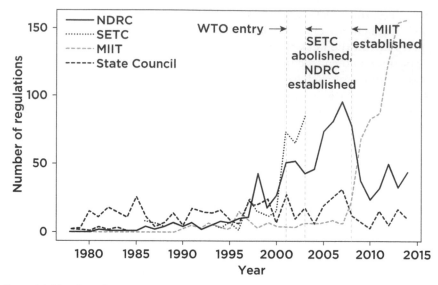

Figure 4.6. Number of sector-specific regulations issued per year by agency
Source: Author's data set.

Implications for the Xi Jinping Era

What has changed since Xi Jinping took over as the CCP's general secretary? As with the Hu-Wen years, the Xi era has been associated with low WTO leverage compared to the years immediately after China's WTO accession. This weak leverage was due not just to China's completion of many protocol commitments but also the inability of WTO members to commit to another round of trade liberalization. China's main source of WTO-related pressure during the Xi years therefore has been in the form of WTO dispute cases.[102] With Xi in charge, the relationship between party and state would appear to have shifted to one of the bureaucracy as agent rather than constituent. In terms of the ability of the party to discipline the state, Xi has emerged as China's most powerful leader since Mao Zedong and led the waging of an anticorruption campaign unprecedented both in the number of cadres who have been purged (including hundreds of high-ranking officials and extending to members of the Central Committee and Politburo and a retired member of the PBSC) and for the duration of its intensity. One source of his political strength might lie in the location of his network, which is largely *disembedded* from the central government. Indeed, Xi has spent almost his entire career—thirty-one out of thirty-eight years prior to being appointed to the PBSC—in various provincial positions (in Shaanxi, Hebei, Fujian, Zhejiang, and Shanghai) and so would have been at no risk of being captured by the central bureaucracy. In contrast, Premier Li Keqiang, who is widely perceived as relatively weak and ineffectual, spent nineteen years of his career in the central bureaucracy versus twelve years in the provinces.

Again using the party leadership's ability to implement administrative restructuring as a gauge of the balance of power between the party and the central bureaucracy, the depth of Xi's administrative reforms announced in the spring of 2018 reflects the uncontested primacy of the party leadership. The plan on administrative restructuring submitted to the National People's Congress for ratification was defined by two major themes. The first was the significant downsizing and streamlining of central agencies. The number of ministries and commissions was cut by eight to twenty-six, while another seven agencies of vice ministerial level were abolished.[103] Notably, the restructuring saw the consolidation of market supervision functions previously fragmented across multiple regulatory agencies, alongside the reduction of policy authority for the developmental agencies. The new State Administration for Market Regulation was created by merging three major regulatory agencies: the SAIC, the SAQSIQ and the State Food and Drug Administration. This new administration also took over policy responsibility for antimonopoly and pricing, functions that had previously been split between the NDRC, MOFCOM, and the SAIC. In the financial sector, the China Insurance Regulatory Commission and the China Banking Regulatory Commission were merged to form the China Banking and Insurance Regulatory Commission.

In contrast, the NDRC (the erstwhile "mini-State Council") saw not just the loss of its authority over antimonopoly supervision but also the removal of several other influential portfolios. Responsibility over climate change was moved to the new Ministry of Ecology and Environment (formerly the Ministry of Environmental Protection), and authority over land and water use was relocated to the new Ministry of Natural Resources. While the substantive consequences of this restructuring will become clearer only over time, these administrative changes potentially strengthen the policy authority of the regulatory agencies while curbing the activist scope of the developmental agencies.

Second and more importantly, this restructuring was aimed at—in the words of an official Xinhua report—"strengthening the overall leadership of the Party" and in terms of the party-state relationship left no doubt as to the ascendancy of the party vis-à-vis the state.[104] The encompassing nature of the party's leadership was encapsulated in another Xinhua report: "Party, government, military, civilian and academic: east, west, south, north and center, the party leads them all."[105] In terms of whether the state acts as an agent for or constituent to the party, then, the administrative restructuring clearly reflected the power of the party leadership and the establishment of "a more disciplined and accountable administration to serve as an *instrument* for Xi Jinping."[106]

Given the combination of a strong party leadership and weak external leverage, the balance of regulatory versus developmental policies on China's economic agenda would depend much more on the preferences of the party leadership, particularly compared to the Hu-Wen years. This is especially so given the unique and precedent-breaking features of Xi's rule thus far. Nevertheless, the argument presented in this chapter provides a useful framework for understanding key differences in the political relationship between the party and its state across different generations of Chinese leaders. Importantly, shifts in this relationship interact with changes in the external leverage to be obtained from international rules to affect the trajectory of economic policies emanating from China's powerful central government.

Reassessing "China, Inc."

Common explanations for the rise of Chinese state capitalism in the latter part of the 2000s have pointed either to differences in leadership preferences between the Jiang and Hu eras or to the diminished allure of market capitalism in the wake of the 2008 global financial crisis. This chapter has shown that neither of those explanations is adequate. Trade liberalization is commonly expected to strengthen market forces, and key insights from historical institutionalism suggest that such policy shifts become entrenched because of increasing returns generated by positive feedback effects. As the beneficiaries of market-enhancing reforms expand, so too does the political constituency for ensuring the maintenance and expansion of liberalizing institutions. This chapter has highlighted the conditions under which such institution building might not occur: when weak agencies use external leverage as a substitute for domestic support, the resulting feedback effects may not generate strong institution-building dynamics. Instead, negative feedback effects become more salient, as the losers from reform find their influence circumscribed. Focusing on the interaction between these domestic bureaucratic politics and external economic rules therefore deepens our understanding of the rise of this state capitalism—not as a product of a coordinated "China, Inc." but rather as the result of politics.

This chapter's focus on the impact of WTO entry on the distribution of power within China's central bureaucracy moreover highlights that liberalization through the WTO creates not just economic beneficiaries and victims in the form of exporters and import competitors but also bureaucratic winners and losers. As the evidence in this chapter underscores, the contest for political advancement between these bureaucratic agencies plays an important role in shaping the policy trajectories emerging from WTO entry, resulting in policy reversals and shifts that cannot be explained by an analysis based on demands for protectionism or liberalization from economic interest groups.

Existing literature has suggested that external commitments by authoritarian regimes might be more credible due to top-down rule and the relative weakness of domestic opposition.[107] This chapter shows, however, that a different set of domestic politics affect the durability of authoritarian commitments. In one-party regimes such as China, the efficacy of relying on external rules as credible commitment to reform depends on the political relationship between the party leadership and the state bureaucracy and on the durability of the leverage to be gained from the external rule. This explanation highlights the importance of connecting the politics of responses to international rules to an interactive relationship between the external environment and bureaucratic politics within authoritarian regimes. While developed democracies tend to have a more stable relationship between political parties and the state that is defined by formal institutions, shifts in the relationship between the political elite and bureaucracy in nondemocratic contexts can have important consequences for the pattern of policy responses to external rules.

While literature on authoritarian durability has focused on the infrastructural power of the state, this chapter has demonstrated the importance of contextualizing this infrastructural power within the broader political relationship between party and state. One benefit of such a framework is that it allows us to observe patterns of politics different from those that usually characterize China. For example, one theory of institutional change in China is that the party-state cycles through periods of centralization, decentralization, and recentralization.[108] This framework has lost its utility in recent years, since the Jiang-Zhu, Hu-Wen, and current Xi eras have all been characterized as periods of centralization. This chapter has shown that the Jiang-Zhu and Hu-Wen eras actually consisted of very different *types* of centralization, thereby raising the possibility of a different type of cyclicality in Chinese politics: that of the state acting as an agent or a constituent of the party.

Chapter 5

FDI and the Quest for National Champions

By the late 2010s, concerns over Chinese state capitalism had co-alesced around an industrial policy issued by the Chinese government in 2015 called "Made in China 2025," aimed at boosting Chinese capabilities in high-tech and intelligent manufacturing. The fears that this policy generated in US firms and government are well encapsulated in one US Senate report, which assessed that the policy "demonstrates that the Chinese government is doing more than merely "breaking the rules"; it is seeking to set new terms for international economic competition."[1] At first glance, the analysis presented thus far goes some way toward supporting these fears of a consolidated Chinese government push to promote internationally competitive national champions: If central policies shifted toward more activist developmental policies after 2006, then the divergence between central and provincial developmental policies should have been ameliorated in China's later World Trade Organization (WTO) period, leading to a coordinated push for market-shaping industrial promotion. I show in this chapter, however, that this was not the case. While the central government has responded to WTO entry with a renewed desire to foster globally competitive national champions across a range of industries, trade liberalization impedes the ability of the central government to achieve this goal. Why is this the case, and when does WTO entry trigger greater policy divergence in some industries rather than others?

To understand the challenges of building national champion industries under conditions of globalization, we need to return to the clash of interests between the central government and its subnational authorities. This tension has been a constant feature of China's political economy, but WTO liberalization exacerbates the problem by enhancing the access of subnational governments to external resources in the form of foreign direct investment (FDI). Even when both central and subnational governments are oriented toward developmental strategies, their divergent interests lead each to deploy FDI toward different types of *political advancement*. The central government, driven by a long-term *regime promotion* imperative, seeks to leverage the technology component embedded in FDI, while the subnational government, driven by a short-term *rank promotion* imperative, seeks to leverage the capital component of FDI. Crucially, then, the question of which state actor holds *contracting authority* over FDI is important, as this authority affects the political purpose that FDI is deployed toward. In industries where contracting authority over FDI is delegated to subnational governments, these authorities are able to deploy FDI toward rank promotion by prioritizing short-term output and employment maximization, bypassing and even undercutting central regime promotion policies emphasizing the deployment of FDI for long-term technological upgrading. Consequently, industries where subnational authorities enjoyed stronger control over FDI experienced greater internal industrial policy incoherence at precisely a time when the center wished to consolidate the positions of its national champions.

This chapter unpacks these dynamics first with comparative case studies of the automotive and semiconductor industries, illustrating the causal mechanisms behind why greater subnational authority over FDI generates policy divergence in strategic industries. I then turn to a quantitative analysis to assess the broader relationship between FDI access and relative developmental activism in central versus subnational policies in China's manufacturing industries. While chapter 3 differentiated between three levels of administration, this chapter simplifies the analysis by comparing the interests of the central versus subnational (i.e., provincial and local) governments. This is because provincial and local government behavior is fairly aligned when it comes to FDI policy.

Explaining Policy Divergence across Industries

The politics of FDI is important to investigate because of the crucial role that foreign capital has played in China's development trajectory and because WTO entry triggered a resurgence of FDI inflows that had stalled during 1998 Asian financial crisis.[2] While trade and FDI can be either complements or substitutes, figure 5.1, showing the net inflow of FDI into China from 1980 to 2006, clearly indicates a complementary relationship. From the figure, we can observe that the first rise in FDI inflows occurs around 1992, when Deng Xiaoping made his famous "Southern Tour" to reignite economic growth after the 1989 Tiananmen crisis. FDI inflows then leveled off and even declined around 1997–1999, increasing again only in 2001 with China's WTO entry.

Existing research on the political economy of FDI has primarily focused on the economic and distributive effects of FDI and factors affecting FDI inflows.[3] One prominent area of investigation examines the link between FDI and economic performance.[4] Some argue that FDI has improved productivity through

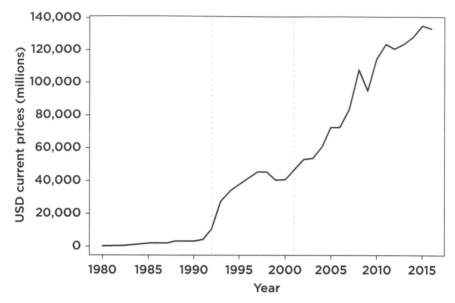

Figure 5.1. FDI flows into China, 1980–2016
Source: UNCTAD.

technological spillovers, while others point out negative effects or identify conditioning variables, such as human capital and financial markets, that mediate economic outcomes. In terms of distributional effects, important scholarship has examined how FDI affects the welfare of elites, of labor versus capital, and why particular types of FDI generate differential sociopolitical consequences.[5] Another large subset of the literature has focused on the question of how to attract FDI, with explanations ranging from the effects of regime type to the design of credible commitments for mitigating the threat of expropriation.[6] The focus therefore has been on how to constrain state behavior to effectively draw in foreign capital, with the implication that FDI, like other external economic forces such as trade, has a disciplining effect on the state.[7]

One important dimension that has been neglected is the question of who within the state holds *contracting authority* over FDI, and the ways in which that affects the political purpose that FDI is deployed toward. This gap stems partially from a tendency to treat the state as a unitary actor and consequently overlooking the divergence of political interests within the state over FDI policy. As scholars have shown, FDI alters politics within the state.[8] This chapter underscores the importance of disaggregating these intrastate interests because FDI comprises a capital component and a technology component. Even when states are able to solve the hold-up problem and successfully attract

FDI, whether the technology or capital component of FDI ends up being prioritized varies depending on the political interests of different substate actors.

I propose that although central and subnational governments are both driven by political advancement, each seeks a different *type* of advancement. The central government leadership is deeply concerned with the continued political durability of the Chinese Communist Party (CCP). The government therefore treats economic growth as a means to *regime promotion* and is focused on raising the long-term viability of the economy. As a result, its deployment of developmental strategies for industries is driven by two major concerns. The first is a view of technological advancement as a fundamental condition for the sustained vitality of the Chinese economy. The central government is therefore interested in harnessing advanced foreign technology made available by WTO entry so as to move the economy up the value-added chain. Additionally, the central government views control over the economy as vital to its long-term political survival. This means securing party assets in industries deemed to have national strategic value due to their technology content and promoting the development of such industries (e.g., automobiles). It further means maintaining ownership over sectors that are strategic due to their status as economic lifelines (e.g., power generation).[9]

In contrast, subnational government leaders treat economic growth as a means to *rank promotion* and seek to maximize short-term growth rates to boost their chances of promotion up the CCP hierarchy.[10] This short-term view of growth means that with WTO entry, subnational governments aim to enact developmental policies in capital-intensive industries that will produce the highest rates of GDP growth in the immediate term. The potential technology benefits that FDI offers are, in relative terms, less important compared to the capital component of FDI. This bias arises partly because developing high-tech innovation capabilities with commercial applications entails higher risks and often longer times to secure profitability and might also generate cross-jurisdictional spillovers that subnational leaders cannot internalize. Moreover, the rank advancement prospects of subnational officials also depend on demonstrated success at job creation and expanding economic activity during their tenures. Hence, subnational states are motivated to enact more developmental strategies in industries that can generate the most employment and short-term output growth in their jurisdiction.[11]

One important implication of this argument is that even when both central and subnational governments are oriented toward enacting developmental strategies to attract FDI, their political incentives generate divergence over whether this FDI serves a technology upgrading or output maximizing purpose. And as numerous subnational governments compete to outdo each other in their bids for FDI, subnational policies end up undermining Beijing's focus on technology upgrading, generating problems in two respects. The first is the problem of excess capacity and duplicative development projects as numerous subnational authorities pursue foreign investment in the same few sectors, replicating each other's efforts. The second problem is that in seeking to maximize short-term growth rates, subnational governments are incentivized to approve foreign investment projects that generate the greatest output in the immediate period, while activities geared toward technology upgrading require a longer time horizon—both from the research and development angle and in terms of the time needed to recruit and train high-skilled labor. Indeed, the problems of excess capacity and poor progress toward technological advancement related to foreign investment are issues that China's leadership has repeatedly sought to address over the years. In a speech at the Central Economic Work Conference in 1998, Zhu Rongji pointed to the tension between the center's emphasis on technology upgrading versus the subnational state's emphasis on short-term growth by noting:

> We must study the markets and keep the overall picture in mind, rather than just consider the interests of one's own locality. I don't want you to engage in joint ventures that produce goods that swamp the market and lead SOEs [state-owned enterprises] in other provinces to collapse. That's why in bringing in foreign investment, if it doesn't really introduce new technologies and new product types, you shouldn't engage in any joint ventures—they won't do much good. I want to stress that enterprises need to pay attention to developing technologies and not simply expand production capacity—they should fund technology development.[12]

Yet cooperation from subnational governments has remained a problem. The need to resolve this tension was still on Zhu's agenda three years later when he emphasized the need to improve the government's ability to rein in investment in his report on the Tenth Five-Year Plan in 2001. In a speech at a national working conference on foreign investment that same year, Zhu noted that

> capital is no longer a prominent problem in economic operations. Hence we should no longer focus on simply seeking scale when attracting investments. Rather, we should work at raising quality and standards when utilizing foreign investment. . . . If we continue to ignore technology and management quality and keep attracting investment just for low quality redundant construction, we will intensify structural contradictions and create greater difficulties and burdens. . . . In order to attract business and investment, some locales are violating central policies, national laws, and regulations. They are following unauthorized preferential policies on tax reductions and exemptions, land use, disposition of state-owned assets, access to industries, and approval procedures. . . . All locales must comprehensively clear up their own policies and methods for attracting foreign investment.[13]

With WTO entry, increased FDI inflows only served to exacerbate center-subnational conflicts over industrial policy. On the one hand, trade liberalization intensified the central government's desire to speed up technology up-grading and foster globally competitive national champions. On the other hand, the central government ends up being thwarted in achieving this goal by precisely the new economic conditions created by WTO entry, as the next section will discuss.

Deploying FDI for Growth: A Comparative Case Study

I use two detailed case studies of the automobile and semiconductor indus-tries to explain why center-local tensions have been deepened by a liberaliza-tion of FDI associated with the WTO era. While the automobile and semicon-ductor industries are different in several ways (e.g., the former is a final product, while the latter is a component), they nevertheless share important character-istics. Both are capital-intensive and require economies of scale in production, and both are high-tech and high value-added industries, with "pillar" or stra-tegic status in the national economy. Specific industrial policy documents to develop each sector have also been released and revised over the years, allow-ing a close comparison of policy trajectories for the two industries. Both in-dustries also experienced significant tariff liberalization as a result of WTO entry, albeit to different degrees.[14]

In putting these case studies together, I relied on official speeches of gov-ernment ministers and industry officials, annual reports from industry year-books that often summarized major events and concerns of the industry, reg-ulations and policies issued by various government agencies, and articles from newspapers and industry-related journals. I further conducted interviews with Chinese and foreign industry managers, journalists on the business beat, industry association representatives, academics, and members of semiofficial think tanks such as the Chinese Academy of Science. Given the national scope of this study, most of the interviews were conducted in Beijing, where there are more industry associations and where firms tend to locate their govern-ment relations managers. Additional interviews were held in Shanghai, Jiangsu, Guangdong, and Yunnan.

Access to foreign technology—and hence FDI—played an important role in the national industrial policies enacted for each industry in China's reform era. The central government deployed the strategy of "exchanging (the do-mestic) market for technology" for both industries in the 1980s, which involved selective approvals of FDI deals conditioned upon the investment taking place through joint ventures between the foreign firm and a selected Chinese "cham-pion" (usually an SOE).[15] The idea was that the foreign investing firm would gain access to China's vast domestic market, and in exchange the Chinese part-ner firm would gain access to advanced foreign technology.[16]

From 2000 onward, however, policy for the two industries diverged. While FDI in the automobile industry continued to be strictly controlled by central government policy emphasizing joint venture requirements and market entry restrictions, foreign investment in the semiconductor industry was liberalized, and contracting authority over FDI was decentralized to subnational governments. Consequently, provincial and local governments focused their semiconductor policies at maximizing the volume of foreign capital entering their jurisdictions so as to meet short-term goals of output and employment creation rather than the central government's goal of technology upgrading. Over the 2000s, the sector saw large increases in wholly foreign-owned enterprises (WFOEs), attracted by favorable tax and credit incentives offered by subnational governments.

Despite the loosening of FDI controls in the semiconductor industry, however, central goals for the two industries remained similar: to develop globally competitive national champions and to close the technology gap with developed economies. Therefore, while auto industrial policy has remained fairly constant, the shift in contracting authority for the semiconductor industry to subnational levels generated substantial divergence in policy focus and dramatic swings in industry governance over the years. In response to subnational policies that had brought in a flood of foreign firm activity without producing significant advances in indigenous technology capabilities, the central government reasserted its control around 2010 through a series of new policies and measures. These initiatives were aimed at recentralizing authority over semiconductor industrial policy, consolidating the industry structure, reducing the dominance of foreign firms, and reorienting policy content toward technological gains for Chinese firms. In contrast, policy for the automobile industry has been relatively more stable, in part due to the central government retaining its contracting authority over foreign investment.

To assess the degree of center-subnational policy divergence for a given sector, I construct a measure of relative policy activism using the textual data introduced in chapter 3. Relative policy activism simply refers to whether the prevalence of developmental language in policies governing a given sector is greater in the central or subnational government. I measure this relative policy activism by taking the simple difference of the strategy prevalence between the central and subnational government policies across industries. In this measure, a positive value implies that the central government placed relatively more developmental emphasis in its regulations for that industry compared to subnational governments, whereas a negative value means that the subnational state placed relatively greater developmental emphasis in its regulations for that industry. Zero values or values close to zero imply little difference between the two levels of government in their developmental emphasis for an industry.

The varying policy trajectories of the two industries can be seen in figure 5.2, showing the relative policy activism in developmental language for central and subnational government regulations in the automobile and semiconductor sectors. The first observable trend is that up until around 2000, developmental

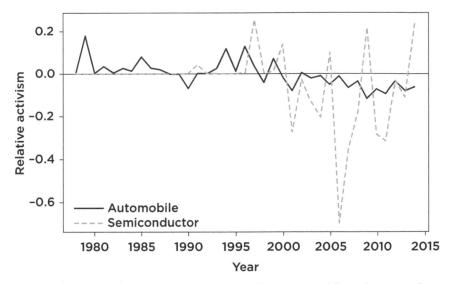

Figure 5.2. Relative activism in developmental policies, automobile and semiconductor industries
Source: Author's data set.

policies for both sectors were relatively dominated by the central government.[17] From 2000 onward, however, trajectories for the two industries diverge. While the overall trend for both is one of greater subnational activism, the relative increase is much greater for the semiconductor industry and more volatile. The semiconductor industry goes through a period of high subnational activism from 2000 to 2008, suggestive of steadily rising fervor from provinces and cities for promoting the industry. From 2009 onward, we observe spikes of greater central activism in later years, indicative of a reassertion of central control. The next sections will explain the relatively stable trend in automobile industrial policies and the reasons behind the swings for policies in the semiconductor industry as well as the repercussions for industrial governance in each sector.

Automobile Industrial Policy: A Consistent Focus on National Champions and Technology Upgrading

In the late 1980s, the central government set out to consolidate the decentralized auto industry and work toward building a few national champions by issuing the "Circular regarding the Strict Control of Passenger Car Production Points," also known as the "Three Large and Three Small" policy.[18] Through a strongly directive approach, six SOEs were selected for national support.[19] The focus on nurturing these selected "champions" was further strengthened through restrictions on market entry for new firms and prohibitions on investments outside of the official production points.[20]

The 1990s saw a series of policy announcements signaling the elevation of automobile industry development to a core national concern. In 1990, the State Council's national industrial policy and Eighth Five-Year Plan (1991–1995) emphasized auto development as a national priority, reemphasizing many of the same policy measures and objectives from the 1980s such as market entry restriction and technology advancement through international cooperation. An official industrial policy for the automobile sector was released in 1994 by the State Council, highlighting its status as a national pillar industry.[21] The document retained strong directive elements by specifying production and enterprise targets. For example, one stated goal was that by 2000, two to three enterprise groups would achieve the status of large enterprises with "considerable capability," while six to seven enterprise groups would become domestic "backbone enterprises." Another target projected that by 2010, three to four large auto enterprise groups and three to four large motorcycle enterprise groups would become globally competitive.

The 1994 policy further established rules on how foreign capital was to be utilized, emphasizing the two core elements of the central government's desire to maximize technology transfer while securing state assets. Measures were set to consolidate the industry and channel resources toward a few chosen state-owned "champions" and to emphasize transferring advanced technology through FDI while protecting the economic viability of its domestic firms. Through the "market for technology" strategy, each foreign firm had to establish a 50-50 equity partnership with a Chinese state-owned firm and was restricted to a maximum of two such partnerships. Under the terms set by the central government, the joint venture also had to establish a research and development unit, utilize technology benchmarked at 1990 international levels, produce primarily for the export market, and prioritize the usage of domestic over foreign parts in the manufacturing process. To further encourage the latter policy, tax incentives were offered to enterprises that could meet specific levels of "indigenization" (i.e., produced by Chinese manufacturers) in their production.[22] This 1994 document, the first industrial policy ever issued specifically for a single industry, was partly effective in that the 1990s saw the establishment of several more joint ventures.[23] Into the late 1990s and up until the eve of China's accession to the WTO, China's industrial policy did not deviate far from the principles of joint venture partnerships for technological upgrading and favoring designated SOE national champions.

In 2004, the National Development and Reform Commission (NDRC) released a new automotive industrial policy to replace the 1994 industrial policy. The focus on fostering a few "backbone enterprises" to become globally competitive national champions did not change and even intensified, as seen by a newly declared goal of fostering enterprise groups that would join the Fortune 500 list by 2010. Joint venture rules similarly remained unchanged. If anything, the 2004 policy heightened the central government's focus on technological advancement by emphasizing indigenous innovation and developing independent intellectual property, creating indigenous brands and stimulating the development of electric vehicles.[24] The Eleventh Five-Year Plan (2006–2011) again emphasized that SOEs would be the "core" for the industry and reiterated the need to raise the competitiveness of SOEs, calling for foreign enterprises to support this overall goal. Despite some small changes in the stance toward private enterprises, the key characteristic of China's automobile industrial policy has been its continuity, with a consistent focus on SOEs as national champions and technology transfer through joint ventures.[25] Importantly, contracting authority for FDI approvals in this industry stayed firmly with the central government, with all new joint ventures requiring authorization by the State Council.[26]

Only in the late 2010s did signs emerge that the "sacred cow" of the joint venture rule would be killed.[27] In April 2017 the Ministry of Industry and Information Technology (MIIT), the NDRC, and Ministry of Science and Technology released the "Medium- to Long-term Plan on the Development of the Automobile Industry," which included a brief line stating that the government would "liberalize the equity ratio restriction for joint ventures in an orderly manner."[28] However, no further details were given on how this plan would be implemented or the expected schedule of liberalization, calling into question when actual changes to the rule would take place if at all. At the same time, a far lengthier section of the policy emphasized the need to "raise the standards of international cooperation," encouraging joint ventures to expand their investments in R&D so as to increase the proportion of domestically developed automobile models as well as encouraging further exchange of technology and human capital between joint venture and domestic firms. Therefore, on balance, while the 2017 policy exhibited hints of relaxation around the joint venture rule, the overall thrust on foreign investment showed little change. If anything, the emphasis on technological gains through access to FDI was strengthened.

It was only in 2018, amid an aggressive trade policy waged by the Donald Trump administration that criticized, among other issues, China's approach of transferring technology via joint ventures, that the Chinese government made more explicit concessions on the joint venture rule. At the Boao Forum held in China in April that year, Xi Jinping announced that the automotive joint venture rule would be relaxed and tariffs on automobiles reduced. A graduated schedule for liberalization was later released: the rule for electric vehicles would be lifted that same year, while that for trucks and commercial vehicles would be abolished in 2020. For automobiles, the joint venture rule would not be altered until 2022, a full four years after Xi's announcement, revealing entrenched opposition to the relinquishing of FDI control over this industry.[29]

Semiconductor Industrial Policy: Divergence and Recentralization

The central government's policies for the semiconductor industry in the 1980s and 1990s strongly resembled those for the automobile sector. Central policies focused on directive measures such as production targets and a reliance on picking favored SOE champions to enter into joint venture partnerships with foreign firms in order to facilitate access to advanced foreign technology. A developmental policy for the semiconductor industry was issued in 1986, informally named the "531 Strategy" to indicate the explicit technology and production targets to universally promote 5-micron integrated circuit technology, have "core" enterprises master the 3-micron technology, and begin to develop 1-micron technology.[30]

A second set of policies was put forward in 1989, focused on dealing with issues of underinvestment and scattered investment, backward technology and weak application links between integrated circuits, and China's traditional industries. Again, the central government's challenge mirrored the automotive industry's problem in the 1980s of overly dispersed production and technological deficiencies. To address these problems, the new central policies aimed to concentrate semiconductor development in two coastal regions: Beijing in the north and the Jiangsu-Shanghai-Zhejiang region in the south. Five "backbone enterprises" were identified for the channeling of state resources (similar to the auto industry's "Three Large and Three Small" champions). In another parallel with automotive industrial policy, three of these five backbone enterprises were set up as joint ventures with foreign firms with the hope that such partnerships would facilitate domestic absorption of foreign technologies.[31]

Continuing with the joint venture strategy, an ambitious development named Project 908 was launched by the central government in 1990 involving a partnership between Huajing Group and Lucent Technologies. While this project ended up mired in delays, a similar effort named Project 909 was launched in 1995. This project aimed to mass-produce .35–.25 micron 9-inch chips, engage in lab and trial production of more advanced .3 micron and .5 micron chips, and to develop capabilities in chip design. Again, a joint venture approach was adopted to try to achieve these technological goals through a partnership between Huahong Group and Japan's NEC Corporation.[32]

Throughout this period, semiconductor policy was therefore largely driven by the central government, with a targeted focus at bringing in semiconductor FDI through joint ventures, an approach very much congruent with the policy for automotive development. The rationale for this was simple: in line with the central government's objective of promoting Chinese economic growth in the long term and pushing the economy up the value-added chain, joint ventures were seen as a vehicle for the transfer of technology from advanced foreign firms, which would allow domestic firms to catch up and move toward the technology frontier faster than they otherwise could. At the same time, equity limits on these partnerships served to protect SOEs from being undermined by foreign competition.

In 2000, the central government's strategy for semiconductor development shifted markedly. The State Council released the "Policy Encouraging the Development of the Software and Integrated Circuit Industries" (also called the "No. 18 document"). The changes enacted in this policy marked a turning point for the semiconductor industry.[33] The policy noted the urgency of developing the industry in light of China's imminent accession to the WTO and highlighted several development goals. The main objectives were to draw capital and human talent into the industry, meet domestic market demand while achieving some degree of exports, and close the research and technology gap with developed economies. Most significantly, the No. 18 document called for drawing in FDI in *all* forms, not just through joint ventures but also WFOEs. These foreign firms would enjoy a range of policy benefits such as exemptions on value-added taxes, tariffs, and customs fees, particularly for large scale investments (over RMB8 billion).

Figure 5.3 shows the corresponding change in the types of firms in the semiconductor industry after the issuance of the No. 18 document. Whereas SOEs and joint ventures were the main forms of registered firms in the late 1990s, the number of WFOEs increased substantially from 2001 onward (particularly in 2003–2004), while the number of SOEs steadily declined. The number of domestic private firms also rose from around 2000 onward, but there is still clearly a large gap between these firms and the WFOEs that came to dominate the semiconductor industry.[34]

In the automobile industry, by contrast, the joint venture rule remained a firm pillar of industrial policy. Figure 5.4 shows that the biggest increase in the type of firm entering the whole-car manufacturing sector post-WTO entry has been in joint ventures, while the number of SOEs has held somewhat steady, declining only marginally compared to the semiconductor industry. There have been limited increases in domestic private firms, while there are also almost no WFOEs in this sector.

Comparing the policy trajectories of the automotive and semiconductor industries, we can observe a striking difference in how the central government has tried to achieve its technology transfer goals. Automotive industrial policy has consistently placed an emphasis on upgrading technological capabilities in Chinese firms through joint venture partnerships, a 50 percent equity cap on foreign investment, and the "exchanging market for technology" strategy. While policy for the semiconductor industry had strongly resembled that for the automotive sector in the 1980s and 1990s, policies for the two industries diverged after 2000 with the liberalization and decentralization of FDI in the semiconductor industry. This liberalization took place at the same time as China's entry into the WTO, resulting in large increases in FDI inflows economy-wide and particularly in the semiconductor industry. However, decentralization did not mean that the central government had given up its goals of fostering national champions in the industry. Rather, the central state hoped that technological advancement in Chinese firms would occur *through* liberalization. Indeed, the No. 18 policy aimed to achieve technology improvement in the

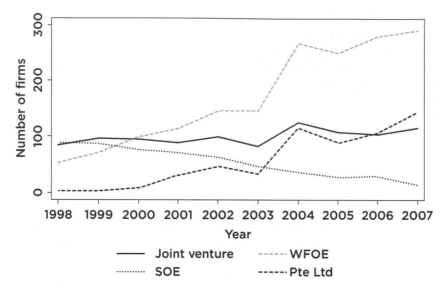

Figure 5.3. Change in types of semiconductor firms over time
Source: National Bureau of Statistics, China.

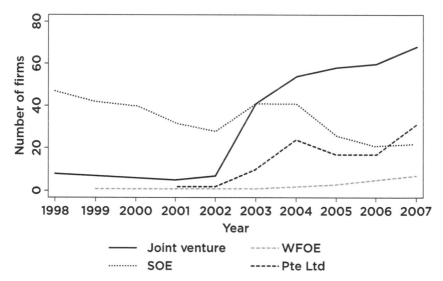

Figure 5.4. Change in types of automobile firms over time
Source: National Bureau of Statistics, China.

industry by specifying additional benefits for investments that brought in advanced technological standards (less than 250 nanometers) and offering tax rebates for firms that reinvested their revenue into R&D activities.

Why did the FDI policy for these two industries diverge? One reason might be the relative dominance of SOEs in each industry around the time of China's WTO entry. Two different interviewees suggested that the joint venture and 50 percent equity limit rules were retained in automobiles and abandoned in semiconductors because there were more centrally owned SOEs in the former industry, which hold a more politically privileged status and have reaped substantial benefits from being in joint venture partnerships with foreign firms.[35] Figures 5.3 and 5.4 also show that SOEs were relatively more dominant in the automotive sector compared to semiconductors in the late 1990s. Therefore, to the extent that central SOEs are seen as core to the central government's interests, or core to the protection of state assets, this might have held the central government back from liberalizing FDI conditions in the automotive industry. Another contributing factor might be the difference between the two industries and the role that semiconductors play as an input into a growing number of manufactured goods. Nonetheless, the central government's core goal of technological advancement in the two industries remained unchanged—except that it was to be achieved through joint ventures in the case of automobiles and through a greater emphasis on explicit tax rebates and other incentives in the case of semiconductors.

Comparing the Consequences of Liberalizing FDI Contracting Authority

China's WTO years saw a great expansion of the semiconductor industry, with revenues increasing from $5 billion in 2000 to $89.3 billion in 2015 and the number of WFOEs rising dramatically, as shown earlier.[36] This growth, however, was far from coherent. While the No. 18 document set the overall direction of industrial policy, individual subnational governments were free to pursue FDI in their own ways, and government policy during this period was characterized by a fair degree of dispersion or decentralization.[37] Geographically, foreign firms were primarily drawn to three clusters in Shanghai, Suzhou, and Wuxi in the Yangtze Delta. Local governments in these areas competed with each other to bring in FDI by offering a slew of benefits ranging from preferential terms on taxation and land use to infrastructure support. Some development zones, such as the Suzhou Industrial Park, were specifically designed to meet the operational needs of semiconductor firms. Therefore the size and shape of each plot of land in the industrial park and the design of the electricity and water supply lines were tailored toward the technical requirements of semiconductor production.[38] In 2007 alone, drawn by such benefits, around two thousand foreign-invested firms located in the Yangtze Delta.[39] Some localities, such as Suzhou in Jiangsu province, became so dominated by foreign enterprises that they were criticized for having turned into a "Western Suzhou."[40]

Why would such a trend contradict or diverge from the central government's goals? As I have argued, the central and subnational governments seek to deploy FDI toward different political purposes. The central government seeks regime promotion via technological upgrading, while the subnational government is motivated by rank promotion via the maximization of foreign capital. Therefore, once FDI in the semiconductor industry was liberalized, subnational governments could use their contracting authority to deploy strategies aimed simply at drawing in capital rather than enacting measures to access advanced technology. The FDI inflows did not automatically generate spillovers of technology to domestic firms for two reasons. First, the simultaneous globalization of production chains during this period meant that many foreign firms shifted their foundry and semiconductor assembly and test services operations in China but retained the technology-intensive parts of the production chain such as R&D and chip design in their home countries.[41] The explosion of foreign investment in semiconductor-related activities in subnational governments was therefore concentrated in just the lower value-added portions of the production chain, while high-tech segments remained outside of China.[42]

Second, FDI can potentially squeeze out domestic firms rather than enhance their access to technology. One study of local semiconductor policies in the Yangtze Delta pointed out that "successful FDI attraction and clustering does not automatically lead to the opportunity and capacity for technological learning and sustainable development by the local indigenous industries." Instead, the authors found that "the major goal of government support for attracting FDI in Suzhou was overwhelmingly at facilitating exports and earning foreign reserves, rather than supporting the development of a domestic semiconductor industry."[43] Therefore, local government success in drawing in FDI was actually counterproductive to the central government's goal of technology transfer. The large share of WFOEs being drawn into Suzhou led to the creation of FDI enclaves with strong networks among foreign firms that were difficult for domestic firms to penetrate. Rather than upgrading local technology capabilities through FDI, the local government's strategy of maximizing foreign capital inflows resulted in the crowding out of domestic enterprises. One interviewee pointed out that FDI dominance had in fact suppressed rather than encouraged entrepreneurship, which was why very little innovation could be seen coming out from an FDI-dominated place such as Shanghai.[44]

That said, this discussion is not meant to suggest that China has failed in its efforts at technology upgrading. There are of course numerous instances of cities and firms that have succeeded in making gains in technological advances. However, these cases of successful domestic upgrading in semiconductor technology have occurred in places where FDI presence is weaker rather than stronger.[45] These dynamics generalize even beyond the semiconductor industry. For example, research has found that domestic technology gains are more likely to occur in cities where foreign firms have a weaker domestic coalitional relationship with local governments.[46]

In sum, the liberalization of FDI to subnational governments in the semiconductor industry meant that the central government found its ability to advance its priorities on technological upgrading to be substantially weakened. While the central government never deviated from its goal of harnessing foreign technology to close the technology gap in this sector, subnational policies were oriented toward short-term goals of maximizing output and employment, resulting in a dramatic policy divergence as relative activism shifted from the central to subnational levels and a stalling of central goals for the industry.

In the automobile sector, by contrast, industrial policy over FDI has stayed fairly constant, and contracting authority over FDI was not liberalized in the same way as in the semiconductor industry. As a result, while the urge for subnational governments to try to maximize the flows of FDI into their localities was very much present in the auto industry, this tendency was restrained by the central government's joint venture rule and restriction on the number of partnerships allowed for each foreign firm. The relative ability of the central state to retain policy control over this sector can be seen in the way in which the central government has repeatedly used its contracting authority over FDI in the automotive sector to meet its national champion goals.

One of the first cases launched against China at the WTO dispute settlement board was over automotive industrial policy. In 2006 the European Union, the United States, and Canada jointly filed a complaint against China on tariffs imposed on imported car parts. The complainants pointed to three policies issued by the central government that they claimed violated multiple General Agreement on Tariffs and Trade articles.[47] Specifically, China imposed tariffs on imports of auto parts equivalent to the tariff on a full vehicle of 25 percent instead of the normal 10 percent if the share of imported auto parts designated to go into a vehicle exceeded a certain threshold.[48] The complainants argued that this policy discriminated against foreign auto parts and amounted to a subsidy of domestic car parts production, raising pressures on foreign parts companies to relocate their production into China so as to avoid the 25 percent tariff. The WTO panel ruled in favor of the complainants, and when China appealed the case, the Appellate Body upheld all but one of the original panel decisions. The Chinese government then complied with the ruling: the NDRC and the MIIT jointly issued a notice to halt the practice, and the relevant Customs decree (No. 125) was also repealed.

While this appeared to be a successful case of foreign companies using international trade law to constrain Chinese industrial policy, subsequent events demonstrated the leverage that the central government was able to wield using its contracting authority over FDI. One interviewee revealed that foreign automotive companies were "punished" after this case, because central government approval was needed when these companies wanted to open new plants in China. Approvals for these new investments were delayed for an extended period, and the central government ended up sending a "very effective message" to these companies about the repercussions of filing complaints at the WTO. The foreign companies involved changed their strategies for operating in China afterward and were no longer so vocal in their complaints about policies that they perceived to be cases of unfair discrimination.[49]

In 2008, the NDRC again used its control over investment approvals to advance its goal of technological advancement for domestic firms in the automotive sector, specifically toward the fostering of new indigenous brands. While no explicit policies were issued to require foreign companies in joint ventures to participate in this indigenization drive, the NDRC had an "unwritten rule" that approvals for investments in new plants would be contingent upon firms submitting a feasibility study or business plan to create and launch a new car model under a new Chinese brand.[50] This strategy was repeated for the production of new-energy vehicles (e.g., electric cars). Again, joint venture firms seeking financing approval for new activities needed to first produce plans or models for new electric vehicles as an unwritten precondition.[51] A report from the US Chamber of Commerce further noted that "foreign automobile manufacturers have been told orally by NDRC officials that approval will not be granted for them to manufacture electric vehicles in China unless they assume a minority stake in a JV [joint venture] with a local company . . . , transfer certain core technology to the JV, and agree to local branding for the vehicles.[52] In each of these cases, the central state used its contracting approval over FDI as leverage to push forward its goal of technology upgrading.

Reassertion of Central Activism in Semiconductors

The surge in subnational policy activism in the semiconductor industry that resulted in a dominance of foreign firms and a lack of clear technological progress in domestic firms was eventually met by a resurgence of central government activism. Significantly, the central government's new policies turned away from FDI promotion toward an emphasis on the advancement of indigenous (i.e., Chinese) enterprises.[53] This change emerged slowly at first. In an effort to reduce the economy's reliance on microchips produced by multinational corporations, the Chinese government set up an association in 2005 to promote the development of smaller local enterprises.[54] The following year a national drive toward "indigenous innovation" was put forward in the "Medium and Long-Term Plan for the Development of Science and Technology (2006–2020)," issued by the State Council.[55] The importance of developing the semiconductor industry was highlighted a few years later in 2012, when the State Council issued its Twelfth Five-Year Development Plan for National Strategic Emerging Industries, which identified seven high-tech and high value-added industries as deserving of state support, ranging from new-energy and high-end equipment to information technology and high performance integrated circuits.[56] These gradual policy movements toward a focus on the promotion of information technology and indigenous innovation were solidified for the semiconductor industry in 2011 when the State Council issued the "Policy to Further Encourage the Development of the Software and Integrated Circuit Industries" (or the "No. 4 document"), the first major revision to official policy since the No. 18 document issued in 2001.

The general thrust of these new policies involved two major components working concurrently: centralization and indigenization. The No. 4 document enumerated a number of tax incentives for the development of larger-scale and more advanced semiconductor plants, reiterating long-standing national policy goals. In other respects, however, this new policy marked a significant departure from existing practices. The most notable change was a shift away from the focus on attracting FDI. Instead, the policy called for improving the domestic credit system so as to increase domestic capital access for enterprises. The policy further encouraged Chinese firms to embrace the "going out" strategy of overseas investments and exports. The one area where foreign resources still remained an aspect of the industrial policy was in human capital, with emphasis placed on attracting people trained overseas back to China. One interviewee suggested that the push for indigenization reflected the central government's growing concern about the large gap between China's consumption and production of microchips, resulting in policies aimed at reducing the country's reliance on foreign-made chips.[57]

This new policy focus was accompanied by greater central government intervention to shape the industry as a whole, a trend observed by more than one interviewee.[58] One component of centralization involved an effort to consolidate state support around a few key national champions. In January 2013, the MIIT announced plans to build major enterprises in the electronics industry via mergers and acquisitions. The stated goal was for China to have established between eight and five major companies with minimum sales revenues of US$16 billion (RMB100 billion) by 2015.[59] An interviewee noted that the consolidation drive was in part motivated by the central government's desire to reduce "unfriendly competition" between subnational governments.[60]

In 2014, the State Council put forward the "Outline for Spurring the Development of the National Integrated Circuit Industry" reiterating the central government's move toward harnessing domestic capital to develop Chinese capabilities in semiconductor production.[61] This document stressed the importance of domestic innovation and warned of the security implications of China's overreliance on foreign imports in this strategic sector. In addition, the document set numerous industry goals to be achieved by 2020, including to have shrunk China's gap with international standards, to have increased domestic sales revenue by over 20 percent, and to be producing 16–14 nanometer chips at scale, among others. In a sign of the central government's more interventionist stance, the Leading Small Group for semiconductor development was also set up. The National Integrated Circuit Investment Fund was announced to support the development of large enterprises in all segments of the production chain. China's major policy banks, from the Export-Import Bank to the China Development Bank, were encouraged to increase their financing to the industry. The need to attract foreign capital, foreign technology, and human talent was relegated to just three sentences at the end of the document.

Later the same year, the government announced the creation of an investment fund of over US$22 billion, part of which would go toward acquisitions.[62] Since then a number of high-profile sales have been made, many with government involvement to ensure the acquisition of the latest technology along the entire semiconductor production chain.[63] Tsinghua Unigroup, for example, through a number of high-profile mergers and acquisitions, had emerged by 2016 as one of the largest semiconductor companies in the country, worth US$2.8 billion.[64] Another change resulting from the focus on indigenous innovation was that government procurement had become a more important policy tool in ensuring that locally made chips had a ready-made market.[65]

The focus on indigenization was further reinforced by the State Council's announcement of its "Made in China 2025" policy in 2015, which listed semiconductor manufacturing as a key component of China's goal of developing domestic strength in high-tech manufacturing.[66] The effects of this increasingly active central policymaking backed by domestic rather than foreign capital did not go unnoticed by industry participants. One interviewee pointed out that from the perspective of some of the foreign companies being bought up, the acquisition drive offered some advantages in adapting to a new policy environment more hostile to foreign enterprises. That is, acquisition by a Chinese firm would offer them greater access to contracts from important Chinese companies such as Huawei that they otherwise would be shut out from. Realizing that "being Chinese" would become increasingly important in the future, they decided to sell to a Chinese company and secure their access to the domestic market.[67]

Another interviewee noted that with government policy now favoring what he called "Red" (i.e., Chinese) enterprises, foreign enterprises had no choice but to find ways to adapt. One option was to cooperate with local enterprises that could provide the "Red hat" to meet official policies favoring Chinese firms. Underneath that hat, the enterprise could have many different "colors" (i.e., other nationalities). Foreign enterprises would have to be more flexible and engage in hybrid arrangements whereby the core of the firm might be Chinese but the technology foreign.[68] In 2015, the CEO of a Dutch firm was quoted in the media adopting just that attitude: "Over the next few years, it's not going to be the same as it is today, where we just ship semiconductors into China. . . . You're going to have to do joint ventures and licenses."[69]

In other words, through the concerted push by the central government to direct domestic capital and consolidation efforts toward national semiconductor champions, firm activity in the industry had effectively swung back toward the joint venture mode of the 1980s. While there was no overt government policy mandating joint venture arrangements, foreign companies were proactively entering into such partnerships to maintain their foothold in China and gain some access to the deep pool of national funds. In recent years Qualcomm, for example, entered into a joint venture project in Guizhou province, while Intel set aside US$3 billion to invest in local mobile microchip firms.[70]

One interviewee suggested that national security concerns were also a driver of the renewed focus on indigenization. Given the military and strategic economic applications of semiconductors, the Chinese government wanted to reduce its reliance on foreign-made chips whose supply could be cut off in the event of a conflict.[71] Those fears shot into national prominence in 2018 when the US Department of Commerce imposed export sanctions on Chinese telecommunications company ZTE for failing to comply with the terms of an earlier settlement regarding ZTE's sale of US-made products to Iran that had violated a US embargo.[72] Although the export ban was later lifted, the scenario of a major Chinese company being crippled due to its access to US-made components being cut off only served to further intensify the policy focus on indigenization that had been building up in previous years. Shortly after the export ban announcement, Xi Jinping delivered a major speech at a national conference on cybersecurity in which he emphasized the need for indigenous innovation and for China to develop greater domestic expertise in core technologies.[73] In May 2019 the Department of Commerce placed another major telecommunications company, Huawei, on an entity list that restricted the exports of US-made components to the company after the Trump administration released an executive order on securing the US telecommunications supply chain.[74] While the exact scope and timing of the export restrictions shifted as the United States and China engaged in a broader negotiation to resolve rising bilateral trade tensions, the US policies on Huawei served to again reinforce China's policy focus on reducing its reliance on foreign firms in the semiconductor industry.[75] These tensions continued to escalate into 2020, which included a new law preventing the federal purchasing of equipment from companies deemed to pose a national security threat, additional restrictions impeding Huawei's access to US technology, and the official designation of Huawei and ZTE as national security threats by the US Federal Communications Commission.[76]

These case studies illustrate the pathways through which WTO-era FDI liberalization intensified the center-subnational policy divergence in the semiconductor industry, driven by the distinct political purposes toward which central and subnational states sought to deploy FDI. As a result, semiconductor industrial policy has been marked by large swings, from a sharp liberalization and decentralization in 2000 with the No. 18 document to a recentralization and indigenization drive about a decade later. The automotive industry, in contrast, has been marked by relatively less policy divergence and steady central activism over time, with a consistent focus on joint ventures and "exchanging market for technology." The ability of subnational governments to subvert central policy in this sector has been weaker (but not absent) compared to the semiconductor industry, as subnational governments were not able to take advantage to the same extent of the increased access to FDI in the WTO era.

Testing Sources of Policy Divergence

This section turns to a quantitative analysis to assess if FDI liberalization has led to greater center-subnational policy divergence in general terms that extend beyond the automotive and semiconductor industries. Overall, we can expect most manufacturing sectors in China to have shifted toward greater subnational developmental activism in the WTO years, given that—as shown in chapter 3—Chinese provinces have on average intensified their developmental strategies in response to trade liberalization. However, the degree to which subnational governments have strengthened their policy activism relative to the central government varies by industry.

I use the measure of relative policy activism described earlier to assess whether developmental policies for a given industry are primarily driven by the central or subnational governments. Figure 5.5 shows change over time in relative policy activism in three industries that are organized very differently: railway equipment, automobiles, and food (fruits, nuts and vegetables) processing. The railway equipment industry was traditionally almost entirely state-owned, and while the share of non-state-owned output in this sector has risen over the years (from just 16 percent in 1998 to 64 percent in 2007), the degree of central control over the sector is still high, since centrally owned SOEs still account for about 32 percent of output.[77] In contrast, the automobile industry has seen slightly more decentralization and greater foreign and private firm presence. While the share of central SOE output in the automobile industry was 17 percent in 1998 (overall SOE output share including that of subnational SOEs was 41%), this figure declined to 6 percent in 2007 (12% for SOEs in general). Finally, the food processing industry has long been highly decentralized and privatized, with less than 0.2 percent of the output deriving from central SOEs (less than 4% SOEs in general) in 1998 and 0.01 percent (0.7% SOEs in general) in 2007. As such, we can observe in figure 5.5 that over time, policies for the food processing industry have become strongly dominated by subnational developmentalism, while developmental policies for the railway equipment industry have remained on the whole relatively more driven by the central government. The automobile industry lies between the two sectors, with developmental policies staying fairly driven by the central government and becoming slightly more driven by subnational governments in the WTO years.

To be clear, it may not necessarily be counterproductive in all cases where subnational governments appear to be taking the lead in driving developmental policy for an industry. If subnational interests are aligned with those of the center or if the center can effectively monitor and credibly punish the sub-

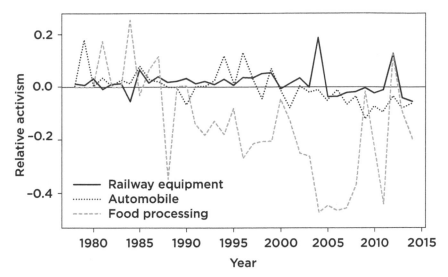

Figure 5.5. Relative activism in developmental policies across three industries
Source: Author's data set.

national government for deviating from its policies, then subnational activism would not be a problem, because the subnational government's activism would be directed toward implementing and fulfilling the goals of the center. As I have emphasized, neither of these conditions are likely in China due to a misalignment of political interests across administrative levels and entrenched monitoring and enforcement problems faced by the central government.

More pertinently, there may be cases where the central government might simply have weak preferences for how a particular industry should be governed and therefore delegated most policy authority over that industry to subnational levels. Even in such weaker cases, however, the competition between subnational governments for investment-fueled growth tends to generate negative spillovers—such as excess capacity—that the central government then has to deal with. The strongest cases for the counterproductive effects of policy divergence apply to the subset of industries that are nationally strategic. In these industries, the central government has strong preferences over industry governance, but subnational policy activism is likely to undermine the central government's policies due to differences between the two in their regime versus rank promotion imperatives. Therefore, across a wide range of manufacturing industries, policy divergence driven by subnational developmentalism has the potential to generate unintended negative consequences for national industry governance.

If the political interests of central and subnational authorities do lead to policy divergence, we should observe the following relationships. First, the central government, seeking to promote industries considered to be nationally strategic so as to secure and enhance the party-state's assets in these key manufacturing industries, is likely to be more active in enacting developmental policies for an industry that is more dominated by central SOEs. Moreover, the central government, seeking to maximize access to foreign technology in order to enhance China's long-term development prospects, is likely to respond to WTO entry with more active developmental policies for an industry that has a higher technology content. In contrast, the subnational government, seeking to maximize output and employment for short-term growth, is likely to be more active in enacting developmental policies for an industry that takes up a large share of output or employment. Additionally, the subnational government, seeking to maximize access to FDI for short-term growth, is likely to respond to WTO entry with more active developmental policies for an industry for which it can attract greater amounts of foreign capital.

I include a number of explanatory variables in the regression analysis to test the above hypotheses. To test if the central government's policy activism is affected by the dominance of central SOEs, I include the industry share of central SOE output as a key variable. I further include the industry share of national output, as subnational governments might be more activist in these industries as a way of maximizing short-term output. As maintaining social stability is important to the performance evaluation of subnational government officials and because this concern feeds into their overall growth strategy, we might expect subnational governments to be more activist for industries that hold a larger share of national employment. I therefore also include the industry share of national employment in the analysis.

To capture the central government's focus on using FDI as a source of technology upgrading, I rely on the Catalogue for the Guidance of Foreign Investment Industries issued by the NDRC, which classifies various industries into different categories of FDI permissiveness, from "Encouraged" to "Allowed," "Restricted," and "Prohibited." Using a new data set quantifying the various investment classifications for all manufacturing industries listed in the catalog, I construct a measure capturing the degree to which foreign investment in a sector is officially encouraged by the central government.[78] The industries for which FDI is most highly encouraged by the central government range from chemicals and shipping to dedicated instruments, aerospace, and medical equipment. The sectors all tend to be technology-intensive, demonstrating the central government's goal of using foreign investment as a means of technology upgrading. A positive coefficient estimate on this variable would indicate more activist central policies for a given industry that holds greater technology intensity.

To measure the degree to which subnational governments have direct access to foreign capital, I construct a measure capturing the foreign share of total capital in an industry for firms registered at subnational levels of government. If the policy behavior of subnational governments is oriented toward maximizing foreign capital, we would expect them to have more active developmental policies in a given industry that has higher subnational shares of foreign capital. I further include the industry share of exports in the estimation, as we might expect subnational governments to adopt more developmental policies for industries that are more export-oriented, given the importance of export-driven growth in China's economy. I include industry-specific tariff levels to capture the exposure of different industries to WTO-led liberalization. As with the analysis in chapter 3, the variable is inverted to improve interpretability: a positive coefficient implies an increase in central policy activism in response to a decline in tariff levels.

Finally, to assess how FDI has differentially affected the responses of central and subnational governments to WTO entry, I include two interaction terms. The first is an interaction term between the tariff and subnational foreign capital share variables, which we would expect to be negative if WTO-led liberalization provokes subnational governments to intensify their efforts to maximize the inflow of foreign capital.[79] The second is an interaction between the tariff and FDI encouragement variables. If WTO-led liberalization intensifies the desire of the central government to attract FDI into technology-intensive sectors, we would expect a positive coefficient for this interaction term.

Results and Findings

Table 5.1 shows the regression results for the analysis estimating the relationship between various industry characteristics and relative policy activism in developmental strategies. A positive coefficient on a variable indicates greater central activism, while a negative coefficient indicates greater subnational activism. Column 1 shows the results from a baseline model without any interaction terms. Column 2 displays results with the addition of the interaction term between tariffs and subnational foreign capital share, while column 3 shows the results with the addition of the interaction term between tariffs and FDI encouragement.

The first notable finding is that there is a positive relationship between an industry's share of central SOE output and greater developmental activism by the central government. This evidence is consistent with the argument that the central government's policy is driven by concerns of regime promotion, specifically to secure ownership and to defend party assets in industries that are nationally strategic. In terms of the relationship between central and subnational government behavior toward FDI, the findings in this analysis are largely supportive of the hypotheses laid out in this chapter. Column 3 of the table shows a positive and significant coefficient on the interaction term between tariff liberalization and central FDI encouragement. This tells us that

Table 5.1 Factors affecting relative policy activism

	(1)	(2)	(3)
Central SOE share	0.15334***	0.15258***	0.15417***
	(−0.05868)	(−0.05414)	(−0.05484)
Output share	0.15509	0.58446	0.657
	(−1.03288)	(−1.13336)	(−1.16424)
Employment share	−2.07945	−1.33671	−1.21371
	(−2.02508)	(−1.99522)	(−1.95698)
Export share	−0.09417	−0.0615	−0.06883
	(−0.11079)	(−0.10492)	(−0.10649)
Foreign enterprise share	0.01716	0.03532	0.01775
	(−0.10734)	(−0.10945)	(−0.10978)
Tariff standard deviation	0.00032	0.00102	0.00006
	(−0.00206)	(−0.00216)	(−0.00213)
Tariff (inverse)	−0.00095	0.00178	−0.00008
	(−0.00177)	(−0.00223)	(−0.00198)
Central FDI encouragement	0.00272	0.00342	0.00356
	(−0.00359)	(−0.00347)	(−0.00335)
Subnational foreign capital share	−0.06164	−0.27649**	−0.30980**
	(−0.11204)	(−0.13953)	(−0.14563)
Tarff (inverse) X subnational foreign capital share		−0.01577*	−0.01728**
		(−0.00821)	(−0.00772)
Tariff (inverse) X FDI encouragement			0.00019**
			(−0.00009)
Year fixed effects	yes	yes	yes
Industry fixed effects	yes	yes	yes
Observations	1,017	1,017	1,017

*p < 0.1, **p < 0.05, ***p < 0.01
Standard errors clustered by industry.
All explanatory variables lagged by one year.

the central government adopts more active developmental policies in response to WTO tariff liberalization as the technology content of a given industry increases. Subnational governments, on the other hand, exhibit a different pattern of policy behavior. Table 5.1 shows first that subnational governments tend to deploy stronger developmental policies as the subnational concentration of foreign capital for a given industry increases. This relationship suggests that part of the rank promotion motive that drives subnational policy behavior manifests itself through short-term maximization of foreign capital. In addition, the negative coefficient on the interaction term between tariff liberalization and the subnational share of foreign capital suggests that subnational governments adopt a more active developmental response to WTO entry for a given industry for which access to foreign capital is greater.

Together, these findings are consistent with the argument that Chinese central and subnational states are driven by different political interests when deploying their developmental strategies. Subnational governments seek to deploy developmental strategies to serve their rank promotion imperative by maximizing short-term growth through an emphasis on foreign capital. The central government's regime promotion imperative leads it to focus on promoting sectors dominated by centrally owned SOEs and technology-intensive industries. The evidence further suggests that while WTO entry increased China's potential access to both foreign capital and technology, central and subnational states sought to deploy these external resources toward different political purposes. The central government directed its WTO policy response toward technology upgrading, while subnational governments focused on maximizing the quantity of foreign capital inflows.

Whither the National Champions?

Since the early 2010s, developed nations have become increasingly concerned that the Chinese government is seeking to export its statist model of development around the world, and in the process undermining the rules of the liberal international order. This chapter highlights the ways in which globalization has in fact amplified the contradictions embedded in China's developmental strategies. While WTO entry heightened the central government's regime promotion emphasis on building globally competitive national champions, weak accountability within the Chinese state created an environment where there was a low probability of punishment by higher-level authorities, allowing subnational governments to pursue their own interests. WTO entry only served to supercharge these center-local differences by strengthening subnational access to external resources in the form of FDI and raising the incentives for subnational governments to bypass central policies in pursuit of political advancement (or rank promotion).

By highlighting the FDI pathway through which WTO entry can affect the divergent deployment of state strategies across industries, this chapter offers a new perspective on the challenges of industrial policymaking under conditions of globalization. Crucially, misaligned political incentives within the state mediate the impact of increased foreign capital inflows on industrial governance. The question of which state actor holds contracting authority over FDI is important, as it affects the political purpose that FDI is deployed to serve, with implications for the overall coherence of industrial policy. The central government, driven by a regime promotion imperative, seeks to leverage the technology component embedded in FDI, while the subnational government, driven by a rank promotion imperative, seeks to leverage the capital component of FDI.

Therefore, rather than restraining state behavior as predicted by existing studies on trade and investment liberalization, the WTO era has been associated with heightened policy tensions within the Chinese state apparatus, at precisely a time when the central government has tried to intensify its industrial policymaking in its quest for national champions.

Conclusion

Reassessing China in the WTO Era

China's entry into the World Trade Organization (WTO) in 2001 represented a historic opportunity to peacefully integrate a rising economic power into the international order based on market-liberal principles. To what extent has this promise of the WTO been fulfilled? For the US government, a clear answer had emerged by the end of 2017: China was not playing by the global trading rules, and the liberal internationalist promise put forward by the WTO had proven false. That year, the United States revised its national security strategy away from a decades-long posture of engagement and toward a more hard-edged treatment of China as a "strategic competitor" whose rise threatened United States national interests.[1] Since then US-China relations have only worsened, with a bruising trade war from 2018 onward that, at the time of this writing, sees no easy conclusion. These bilateral relations further deteriorated in the spring of 2020 with the emergence of the COVID-19 pandemic, and whether the damage can be repaired remains to be seen.

While the US shift in national security position back in 2017 was driven by military as well as economic considerations, the economic and trade-related actions that the US government subsequently took against China were grounded in three related beliefs. First, engagement through the WTO was not working. Second, China was instead strengthening its own economic model based on what has come to be termed "state capitalism." And third, this state capitalist model was being exported globally through national state-owned champions and an assertive industrial policy, thereby threatening the existing structure of international economic cooperation.

The analysis presented in this book forces us to question all three beliefs. This study has emphasized that despite over four decades of nearly uninterrupted economic growth, there is little consensus within the Chinese party-state over how best to govern the economy. Instead, China's policymaking process is beset by intense contestation over what measures and policies are most appropriate to keep its economic engine going. The policies that we observe being adopted by the government in response to WTO entry are therefore in part a product of these internal politics rather than the expression of a single internally coherent Chinese economic model. Nor can these policy responses be accurately characterized in the dichotomous terms of state versus market, protection versus liberalization, or compliance versus defiance. Instead, this book highlights the broader *range* of responses with which the Chinese state—and the many actors that comprise it—actively engaged the forces of globalization. In broad terms, these responses are conceptualized in terms of three competing state strategies: a directive strategy that is market-*substituting* in nature, a developmental strategy that is market-*shaping*, and a regulatory strategy founded on a market-*enhancing* mode of governance.

A characterization of Chinese economic governance as one that is fundamentally at odds with the rules of the WTO therefore misrepresents the actual policy trajectories that emerged from China after accession, which were heterogenous across multiple dimensions of the complex Chinese governing structure and also shifted over time. In explaining these policy trajectories, this study departs from analytical treatments of states as unitary actors and frameworks that implicitly assume that authoritarian regimes operate by top-down command. Instead, this study unpacks the authoritarian Chinese state into its multiple actors who wield authority over economic policy and delineates the channels through which these actors engage with the new challenges and opportunities introduced by globalization. Rather than constraining or disciplining the state, the WTO's binding rules and liberalizing forces were filtered through a domestic governance structure that is simultaneously fragmented and highly competitive, allowing actors within the state much room for maneuver and wide scope to craft their policy responses.

While the analytical scope of this study is broad and cross-cutting, the explanation that it puts forward is simple: Chinese political-bureaucratic actors adapt nimbly to the pressures and opportunities of external liberalization and adjust their strategies to advance their political interests. Their responses therefore shift, depending first on their likelihood of being punished, which falls unevenly within China's decentralized and authoritarian governing structure. Depending on where one sits in the sprawling party-state, the threat of sanction might come from the WTO, the party leadership, or the central government. Second, their responses are also shaped by whether the economic and bureaucratic dimensions of WTO entry enhance or threaten their prospects of political advancement. The economic dimension involves the threat of import competition versus export opportunities and enhanced access to foreign capital, while the bureaucratic channel comprises new trade rules that expand the policy influence of some actors while curbing that for others. Consequently, Chinese substate actors adopted economic strategies in response to WTO entry that diverged not just across administrative levels but also over time, within the central government and across industries.

China's hierarchical administrative structure meant that instead of a convergence toward liberalization, we observed distinct strategic responses to WTO entry from governments at different levels of authority. Greater WTO accountability provoked a regulatory response from the central government that corresponded to governance norms promoted by the WTO. At the same time, accountability to the WTO weakens at lower levels of government. Subnational provincial and local governments that were relatively shielded from the WTO's bureaucratic demands adopted largely developmental and directive strategies, respectively, to adapt to new economic conditions of liberalization. More industrially diversified provinces strengthened their developmental policies to seize export opportunities associated with WTO entry, while poorly diversified prefectures and counties, on average, enacted more directive policies to mitigate the dislocating effects of increased import competition and defend growth rates in their jurisdictions. Analyses focused on whether the WTO was a "success" or "failure" in transforming Chinese economic governance therefore overlook the important ways in which the WTO had varied effects on the political incentives of different actors within the state, generating a stark policy divergence.

Nor have these trajectories remained stable over time. This book provides a new explanation for the emergence of what has come to be called Chinese "state capitalism." The central government's developmental turn toward more activist industrial policy cannot be satisfactorily explained by popular accounts, which point to either the 2008 financial crisis or the preferences of the Hu and Wen leadership. Rather, the policy shift rested once again on an interaction of external and domestic factors. Externally, the leverage that could be derived from China's WTO commitments weakened over time, thereby reducing the ability of regulatory agencies to use international rules to advance their policy agenda. Domestically, the degree to which the Chinese Communist Party (CCP) leadership could push for regulatory reform depended on its political relationship with the central bureaucracy. During the Jiang Zemin–Zhu Rongji era, the leadership was able to effectively discipline central agencies and commit the bureaucracy toward regulatory reform because their network ties were largely *outside* the central state. In contrast, in the Hu Jintao–Wen Jiabao period, the leadership could not punish central agencies without having those penalties fall on members of their own network within the bureaucracy; members to whom they were reciprocally accountable and whom they relied on as a key political constituency. This shift in party-state relations, combined with the weakening of WTO leverage, empowered industrial policy agencies within the central government with greater influence over the direction of economic policy, producing a shift in central trajectory toward active industrial policymaking.

This emergence of state capitalism, however, does not translate into a more coherent strategy of promoting national champions. Instead, the global integration brought about by WTO entry has intensified the conflict between central and subnational governments over the governance of key industries. While US trade policy under the Trump administration has emphasized the threat from Chinese policies such as the "Made in China 2025" plan discussed in chapter 5, such concerns rest on the assumption that the policies issued by the central government will be faithfully carried out by subnational authorities. Globalization, however, has complicated this policy process. While WTO entry heightened the central government's desire to build globally competitive national champions, paradoxically it has also enhanced the ability of subnational governments to bypass the center by increasing their access to foreign capital. The central government's national champions policy is therefore more likely to be undermined by divergent policies from subnational governments in industries where contracting authority for foreign direct investment (FDI) has been delegated downward. This distortion of central policies occurs because subnational authorities are politically motivated to deploy FDI toward rank-promoting purposes of short-term output and employment maximization rather than toward central regime-promotion concerns with long-term technological upgrading.

In sum, this book offers a new explanation for why, rather than constraining authoritarian states or credibly committing them to global economic rules, WTO entry alters politics within the state and provokes divergent responses from different substate actors in ways that do not necessarily strengthen the transition toward market economy governance.

Implications for the Domestic Politics of Globalization

The current rules of globalization are founded largely on market economy principles common to developed democracies. As free trade has expanded, these rules have moved beyond a focus on border barriers such as tariffs and quotas to encompass increasingly demanding and intrusive requirements about the content of domestic policies. For today's nation-states, globalization therefore requires not just a lowering of borders but also far-reaching modifications to one's domestic economic institutions. The political challenge that this type of deep integration poses is amplified by the growing diversity of states that are participating in trade agreements. Countries ranging from Chile to Vietnam and Brunei are all signatories to the megaregional Comprehensive and Progressive Agreement for Trans-Pacific Partnership (CPTPP), to name just one prominent example. Analyses of the domestic politics of globalization therefore need to place far more focus on the ways in which the state—and the diversity of actors that make up state structures—respond to the economic and bureaucratic demands of external integration.

The first contribution that this research makes is to move beyond the framing of responses to international rules as an either-or choice between compliance or defiance, or protection versus liberalization. This book makes clear that such framings obscure the full range of strategies that states can adopt in engaging with the economic and bureaucratic effects of international trade agreements. The state strategies investigated here are distinct from approaches that conceive of state governance in quantitative terms of high versus low levels of intervention. Rather, each strategy rests on a specific modality of state-market relations: market-enhancing, market-shaping, and market-substituting.

Of course, this is not the first study to examine variation in modes of economic governance. Peter Evans proposes four different roles that the state can play in developing the economy and demonstrates these differences in comparisons of the information technology industry in India, Brazil, and Korea.[2] The varieties of capitalism literature has emphasized that even among developed capitalist societies, market economies can be organized along two different logics of "liberalism" and "coordination."[3] Kathleen Thelen emphasizes that the trend of liberalization can manifest itself in different trajectories across developed democracies, undergirded in each case by different political coalitions.[4] Other scholars have noted the multiplicity of ways in which states can engage with globalization.[5] All of these studies, however, emphasize differences in governance strategies *across* states. In contrast, this book focuses on the pathways through which trade rules and economic liberalization reconfigure politics *within* the state and explains why different substate actors adopt distinct strategies to adjust to the new economic and regulatory demands of the WTO.

By placing the analytical spotlight on China's vast fragmented governance structure and the complex cast of actors that make up the party-state, this study highlights that the state and its component authorities are far from disinterested actors in the globalization process. Because the rules of globalization increasingly encroach upon issue areas traditionally restricted to domestic policymaking, trade liberalization has come to create winners and losers not just in the economy and society but also in the bureaucracy. This book therefore departs from analyses that focus on how political actors—from legislators to the executive—serve primarily to aggregate or balance between the interests of domestic groups and societal welfare at large and in which the bureaucracy is largely an omitted variable.

Rather, as scholars such as Judith Goldstein and Gregory Shaffer have emphasized, the interests of the bureaucracy are directly implicated in the politics of globalization.[6] WTO entry affects the advancement prospects of various substate actors—subnational governments and central agencies—differentially. For some, WTO entry brings the opportunity of more resources (export markets, foreign capital, or enhanced policy influence), while for others it presents a threat (via import competition or curbs on bureaucratic discretion). In joining the WTO, China also opened itself to the possibility of formal sanction and retaliation from violating WTO rules via the dispute settlement mechanism as well as reputational costs and informal pressures emerging from WTO processes such as the transitional review mechanism. However, the probability of being sanctioned by the WTO varies unevenly and over time for different actors within the colossal Chinese state. As a result, WTO entry does not lead to a uniform policy response. Instead, the common shock of WTO entry triggers varying logics of adjustment within the state as different substate actors adopt responses best suited to advancing their particular political interests under new economic and bureaucratic conditions.

This disaggregation sheds light on why various assessments of the impact of WTO entry on Chinese economic policy seem to be so at odds with one another. Different studies zoom in on specific aspects of or agencies within China's governing apparatus, leading each to emerge with its own set of conclusions. For example, a number of studies examining China's behavior within the WTO observe that the Chinese government has not only largely fulfilled its WTO commitments but has also become highly adept at deploying WTO rules toward its own advantage—in other words, that China has become a status quo rather than revisionist player in the multilateral trading system.[7] However, these studies draw their inferences primarily from the behavior of the Ministry of Commerce, which represents China in Geneva and for which the costs of WTO violation are far larger compared to the costs faced by other actors in the Chinese government. At the same time, contrasting analyses pointing to Chinese domestic industrial policies as examples of WTO-inconsistent behavior base their evidence on the behavior of the National Development and Reform Commission (NDRC), a powerful agency within the central government that has little direct interaction with the WTO.

By applying quantitative textual analysis to a corpus of industry regulations issued by a multitude of governmental units, this study is able to bridge the contradictory findings of different analyses focused on a smaller set of industries or agencies. In doing so, the study strikes a careful balance between macro analyses that obscure subnational diversity and local-level analyses that face constraints with generalizability. The textual analysis further offers a new set of measures for understanding the contestation within the multilayered Chinese state over economic governance and for identifying shifts and divergences in policy trajectories. The data and methodology therefore add to the ways in which we can study the complexity of politics within the Chinese bureaucratic system, which encompasses multiple subsystems of interests and networks working sometimes in tandem but more often than not at cross-purposes.

This research additionally demonstrates the unexpected pathways through which domestic politics can frustrate efforts by national leaders seeking to use international commitments as leverage for domestic reform. China's WTO strategy of using external rules to drive domestic reform is a familiar one that many countries at different stages of development have deployed, from Mexico's membership in the General Agreement on Tariffs and Trade and subsequent signing of the North American Free Trade Agreement (NAFTA) to Central and East European nations seeking to join the European Union (EU) and Japan's commitments to a range of domestic reforms under *gaiatsu* (foreign pressure) from the United States in the late 1980s and early 1990s.[8]

What we have seen from the case of China is that the efficacy of relying on external commitments to drive domestic reform is contingent on additional factors. The first relates to the depth and permanence of the commitment itself. China's WTO commitments were undoubtedly substantial, as earlier chapters have indicated. However, entry into the WTO arguably does not lead to a permanent loss of authority over policy in the same way that monetary union (e.g., Eurozone membership) eliminates control over interest rates. Nor do these commitments match up to the deep political and institutional reforms required to join, for example, the EU.[9] Instead, most policy actions become more deeply contested in domestic politics and subject to challenge and sanction at the WTO's dispute settlement board.

A second reason that the strategy of deploying external rules for domestic leverage may not be durable is that long-term support for policy reform depends on the strength of domestic coalitions for or against a particular change. The potential leverage that reform-oriented agencies could extract from WTO rules was strongest in the years immediately following China's entry due to the highly specific timetable of commitments laid out in the country's protocol of accession. The implementation of these commitments was further closely monitored by China's trading partners via the transition review mechanism. However, most of these commitments were scheduled to be completed by 2005–2006, leading to a substantial weakening of WTO-related leverage over time. In such cases, where external leverage is effective only in the initial stage to overcoming domestic opposition, the durability of policy change depends on the degree to which there are positive feedback effects or linkages with new political and social groups such that coalitions of domestic support emerge to entrench the policy change.[10]

The importance of these linkages and feedback effects has been borne out in cases outside of China. Scholars writing about why there had been less backsliding of political reforms in new EU members from Central and Eastern Europe than one might otherwise expect (at least around 2007 or so) point to the importance of two factors: financial incentives to stay on track derived from conditional EU funding and the expansion of human and economic linkages with the rest of the EU.[11] These linkages not only generated peer pressure among government officials to keep to EU norms but also created processes of socialization at the elite level, generating a positive feedback loop to sustain institutional change.

Why might it be that regulatory reform in China was not supported more strongly by domestic beneficiaries that, with WTO liberalization, might have generated similar feedback effects to entrench the momentum of reform? One might imagine that exporters and consumers would be two strong domestic interest groups supportive of regulatory reform, given their interests in expanded access to world markets and cheaper products at home. In the case of consumers, it could be that the challenge of collective action renders this group ineffective as an interest group, particularly under authoritarian conditions.

As for exporters, it is uncontroversial that they stand to benefit from trade liberalization relative to protectionism. However, it is not clear that they would prefer regulatory policies aimed at leveling the playing field to developmental ones, since the latter are likely to contain benefits, such as subsidies and credit incentives, that discriminate in favor of their industries.

In terms of positive feedback effects within the bureaucracy, similar dynamics in China may have been narrower due to the technocratic nature of trade policy. Socialization into the WTO "club" occurred almost exclusively between the Ministry of Commerce and their counterparts in other trade ministries as well as the WTO Secretariat.[12] Other important economic agencies, such as the NDRC, were not systematically involved in WTO-related policy processes and remained hostile to the international organization. Therefore, where external leverage serves as a substitute for rather than a stimulant to the development of domestic support and complementary linkages, the effectiveness of a change in policy is likely to fade over time, opening up space for opposition or losers from reform to advance their own agenda. As this book has emphasized, these reform losers do not fade away and can mobilize at a later date to fight back against liberalizing forces. Similar dynamics have been documented in the case of Vietnam's WTO-era policy trajectories. There, WTO entry galvanized the conservatives in the government to restructure state-owned enterprises (SOEs) into state enterprise groups that might better withstand the onslaught of foreign competition, thereby strengthening—rather than weakening—the state sector in Vietnam.[13]

The Political Economy of a Globalizing China

China's entry into the WTO was not only a major turning point for the international trading system but has also had deep implications for its domestic economic governance. While the fragmented authoritarianism framework has long been used to explain the contradictions and opposing forces at work in Chinese policy processes, this book connects the politics of globalization to these domestic tensions and draws out the ways in which WTO entry has altered the interests and actions of various Chinese bureaucratic actors across various dimensions of the party-state.[14] Rather than economic competition and global rules exerting a disciplining effect on the Chinese state, WTO entry exacerbated long-standing tensions between central and subnational governments. Importantly, while much of the literature has emphasized the ways in which fragmented authoritarianism leads to poor and conflicted policy implementation, this book points instead to how China's fragmented domestic structure mediates the forces of globalization, leading to policy divergence.

Additionally, this study emphasizes that WTO entry has intensified the contestation for policy influence between central government agencies. By examining the relationship between external rules and the powerful central bureaucracy in Beijing, this analysis reveals a new dimension to the politics of networks in China. Existing studies have been oriented toward examining how networks structure political competition and policymaking or how membership in networks affects political advancement.[15] As a result, these studies treat members within a network (or faction) as sharing a common interest. In contrast, I focus on conditions under which interests can diverge within a network, in particular between the party leadership and the central bureaucracy. Indeed, we cannot assume that the central bureaucracy will always act as a loyal agent to its political leadership even in a single-party regime. Under circumstances where leadership and bureaucratic interests are misaligned, the question of where the leader's networks are embedded comes to play a significant role. And while studies have emphasized the role of informal networks in facilitating policymaking, I focus on the role that network embeddedness plays in shifting the locus of elite policymaking (between the party leadership and the central state). In so doing, this analysis brings attention to an important channel through which politics within networks can hinder, rather than facilitate, the party leadership's ability to get its policies implemented.

The findings further emphasize a new dimension to how accountability operates in China by highlighting the counterintuitive role that network embeddedness plays in determining policy outcomes. In her influential study, Lily Tsai suggests that one factor affecting public goods provision in Chinese villages is whether local officials are embedded in local solidary groups.[16] Specifically, she argues that embeddedness improves public goods provision by raising the social obligations of local officials to contribute to village welfare, generating an informal channel of accountability in the absence of an electoral mechanism. This study demonstrates that at the top echelons of governance, far from the village, embeddedness can in fact *undermine* accountability. When the networks of party leaders are deeply embedded in the central bureaucracy, accountability channels are broken because of the politics of reciprocity that permeate a network: the leadership cannot effectively punish the bureaucracy for deviating from its goals without also inflicting that punishment on members of its own network. As such, the period associated with more assertive state capitalism in China has occurred during the tenure of leaders whose constituents were deeply embedded in the central state.

While much literature on the political economy of China has focused on institutional change, there has been little work that systematically brings in the impact of China's growing participation in international institutions. As China's economy becomes increasingly globalized, it has become more important for research to incorporate the external dimension directly into the analysis.[17] Although various studies all point to WTO accession as having some effect in driving the emergence of a regulatory state, the broader theoretical implications for how globalization has affected the patterns and sources of institutional change in China remain relatively unexplored.[18] One study sees WTO-era liberalization as having led to a dualistic industrial governance regime according to a "strategic value logic," with a relaxation of controls over industries with low strategic value and a tightening for strategic industries.[19] In contrast, other works see globalization as either providing a bottom-up process of change bringing Chinese institutions into convergence with those in developed democracies or sparking a more top-down process of transformation, driven by interactions at the international level.[20]

Other theories of institutional change in China have been by and large focused on domestic factors and can be broadly divided between gradualism and experimentation on the one hand and cyclical patterns and great reversals on the other, mirroring to some extent tensions in the institutions literature between endogenous change and "punctuated equilibrium."[21] Scholars writing from the gradualism perspective see China's economic reform process as evolving from piecemeal initiatives to more coherent macroeconomic reforms, as market rationality incrementally expanded to different spheres of the economy.[22] Other authors argue that China's unique approach of "experimentation under hierarchy" and "directed improvisation" allows the center to spur national reform through local innovation and by adaptively building on weak institutions.[23] Scholars adopting a cyclical view of change in China see the party as oscillating between periods of relative control and relaxation.[24] For both gradualists and experimentalists, China's WTO entry marks a move toward greater rule-based governance, with WTO rules imposing a common set of principles for governing the whole economy and even reducing the scope for policy experimentation.[25] Scholars writing in the cyclical tradition might see WTO entry as a centralizing moment and expect future periods to bring about more decentralization (e.g., with localities engaging directly in global trade).

This book, in contrast, offers a theory that explains why China's domestic governance structure generates varying politics of adjustment to the common shock of WTO entry, leading to policy responses that are neither linear nor cyclical, not top-down and also not bottom-up. Instead, the pattern of change shifts, depending on which dimension of China's complex party structure is being examined. Rather than bringing greater harmonization to Chinese economic policies, the forces of WTO entry were filtered through China's decentralized state, provoking central and subnational actors to intensify their economic strategies but along separate paths. Central policies also shifted over time as WTO leverage reconfigured the distribution of power between economic agencies, leading to regulatory strengthening in the initial WTO years followed by the rise of industrial policy in later years. And instead of a strategic value logic guiding the degree of state control over different industries, policy conflict has increased, even within strategic industries such as semiconductors, as subnational governments leveraged external resources to bypass central policies promoting national champions. The resulting pattern of institutional change is therefore multidimensional, reflective of the numerous subsystems of power relations contained within China's massive governance structure.

China and the WTO: A *Sui Generis* Case?

Can these findings—and the overall theoretical framework—be applied beyond China or in other policy domains? In terms of issue area, the theory is likely to extend to other decentralized areas such as social policy but less so to centralized matters such as national security.[26] Apart from its authoritarian regime type, the main features of the Chinese case that may limit generalizability are the size of its state and its domestic governance structure, where quasi-federalism and weak enforcement generate scope for subnational actors to behave with substantial autonomy. In contrast, small or highly centralized states are likely to have stronger internal coordination over policy responses. To some extent, then, the findings presented in this book may reflect dynamics unique to China. However, the scope of this research can and should extend beyond the Chinese case.

While China's domestic economic reform process could be considered unique for its heterodox incremental and experimental nature, the resulting internal bureaucratic tensions between different approaches to economic governance is something common to many countries. And although China's protocol of accession was more extensive than what other newly acceding WTO members have had to sign on to, the depth of adjustment that China had to undertake can be considered an early case of the trend toward ever deeper terms of integration in trade agreements. The rules governing SOEs in the CPTPP, for example, go beyond even what China committed to in the WTO.[27] And importantly, while China's governance structure is complex, the hierarchical policy dynamics stemming from China's decentralized structure is by no means unique and is of significance to a wide range of other large federalized countries. Finally, the growing diversity of countries joining the WTO and trade initiatives such as the CPTPP raises the salience of understanding how nondemocratic regimes respond to deep-integration agreements.

Competing State Strategies

First, should we expect other countries to respond to global integration by drawing from the same typology of regulatory, developmental, and directive state strategies? Developing countries such as Vietnam and other communist or postcommunist regimes transitioning from the planned economy are more likely to have the same kinds of internal contestation between these three state strategies. Other systems without a legacy of economic planning might find that political and policy differences are mainly divided between developmental and regulatory strategies. Nonetheless, disagreements over the role of the state in economic governance exist in all modern societies, and measures geared toward market-shaping developmentalism can be found even in countries where they might be least likely to be enacted, such as in developed market economies.

While the United States is often held up as an example of a highly liberal and competitive market economy, US developmental policy nonetheless has a long history dating back to Alexander Hamilton's 1791 *Report on the Subject of Manufactures* advocating the deployment of subsidies and tariffs for industrial development. Moreover, this orientation toward market-shaping governance has persisted over time in the US economy. Scholars point out that at the federal level, developmental policies can be found in the activities of agencies charged with scientific advancement such as the Defense Advanced Research Projects Agency, the National Institute of Standards and Technology, and the National Science Foundation.[28] At the subnational level, the initiatives launched by state governments not only extend state support to the promotion of high-tech industries or depressed regions but also set targets in areas such as job creation. Amid global competition for investment, these developmental efforts have only increased over time. The Michigan state government, for example, in 2016 provided a grant to set up the Michigan-China Innovation Center with a goal to "create jobs, attract investment, and increase competitiveness in Michigan by developing win-win ventures with China."[29] The center partners with a host of organizations within the Michigan state government and not only advertises Michigan's low corporate tax rate of 6 percent but also that the state government "provides millions of dollars in support each year for business expansions and growth through its resources, incentives, and loans." None of these activities would look out of place in a Chinese province.[30]

We can observe similar heterogeneity in economic governance approaches within the EU. The 1986 Single European Act made the establishment of a single market one of the EU's main objectives, requiring the breaking down of intra-EU market barriers, establishing a common set of rules and standards for competition, and eliminating potential cross-national distortions such as subsidies. This liberalizing orientation has met with resistance not just from member-states whose domestic political economies are grounded in other economic strategies (from Germany's system of coordination to French dirigisme) but also from various agencies within the EU bureaucracy itself. For example, different European Commission (EC) agencies held distinct views on the best approach to boost European competitiveness via the 2000 Lisbon strategy. The Directorate-General for Enterprise and Industry preferred a more market-shaping approach to create "European champions," while the Directorate-General for Competition advocated more market-enhancing measures aimed at strengthening the level playing field for firms.[31]

Even amid liberalizing efforts to create a single supranational market, the EU has launched developmental policies aimed at boosting European competitiveness. For example, the "Europe 2020" strategy involves substantial funding (450 billion euros across five "structural and investment funds") aimed at issues ranging from innovation and technology to small- and medium-enterprise growth.[32] Furthermore, nation-specific differences over the appropriate form of state intervention persist, as seen in the issue of state aid to firms and industries. In 2003, Germany, France, and the United Kingdom led a trilateral critique calling for a reexamination of the EC's policy on state aid, questioning the liberalizing assumption that enhancing competitiveness at the EU level is the best way to promote growth. In 2016, the debate over state aid was brought into sharp relief when the EC decided that Ireland had illegally provided Apple with 14 billion euros of state aid, a ruling that Ireland subsequently fought in the EU's General Court. While Ireland has won this court case, conflict between the EC and European nations over the legitimate scope and modality of government intervention is likely to continue.[33]

In short, there is no consensus, even in countries with highly developed market economies, over the appropriate role of the state in economic governance. The key distinguishing feature in the competition between state strategies in the Chinese case versus other countries therefore more likely lies in the relative dominance of each strategy within a country's domestic political economy and in the specific political processes through which these differences are worked out rather than any absence of contestation.

Federalism and the Challenge of Globalization

This challenge of mediating disparate and conflicting domestic approaches to economic governance is arguably more pronounced in large decentralized countries, because differences in economic strategies between central and subnational authorities tend to be sharper, while the ability of the central government to sanction its lower-level governments tends to be circumscribed. This means that large decentralized polities face a common challenge when making policy under conditions of globalization—that of reconciling their international obligations with misaligned preferences at subnational levels. Importantly, these hierarchical tensions extend across a wide variety of regime types.

For example, literature on US trade policy has examined the differing trade preferences of officials at higher versus lower levels of the US federal system. In broad terms, politicians from smaller constituencies tend to face stronger demands for protection relative to those who oversee larger constituencies. As a result, the US president (who has the largest constituency size) is less protectionist relative to Congress, while the Senate is less protectionist relative to the House of Representatives.[34] These preferences show up in the more developmental strategies of subnational states in the United States that, like their Chinese counterparts, are engaged in intense competition to attract firm activity and investment to their localities. Similarly, while the German federal government has been characterized as largely rules-based and regulatory (*Ordnungspolitik*), eschewing targeted support or central planning, German regional governments exhibit much greater engagement in market-shaping activities. For example, the Saxony government's funding to support Volkswagen investments in its region was ruled illegal under EU rules on state aid.[35] This literature further resonates with scholarship on the politics of decentralization in Latin America (in particular Brazil and Mexico) and India, emphasizing the sources of subnational variation in economic policies.[36]

Globalization, then, has the potential to exacerbate these federal-subnational differences by triggering a *divergence* in policy trajectories, as seen in the case of China. For instance, one study finds that access to FDI emboldens Vietnamese provinces to engage in policy "fence-breaking," pursuing economic initiatives lying outside of central directives.[37] This research echoes the center-local dynamics discussed in chapter 5, underscoring the potential importance of extending research on FDI and contracting authority to other contexts. Analyzing political preferences rather than economic policy, scholars have found that the increased import competition in the United States generated by China's accession to the WTO has caused greater political polarization in localities more exposed to this "China shock."[38] This political polarization may then translate into greater divergences in local policies.

These divergent policies might be more easily managed if central authorities had strong domestic sanctioning authority over their subnational units. However, while the authority to negotiate and represent a country in trade matters is commonly situated at the national level, in most decentralized countries subnational authorities enjoy a high degree of autonomy over economic policies. As a result, as we have seen with the case of China, lower-level authorities are somewhat insulated from the costs of violating international trade commitments. And indeed, this dynamic is not unique to China. The policy conflicts that globalization generates in decentralized polities are common across regime types and stages of development, extending even to cases that are *least likely* to resemble China such as with the United States, the EU, and Canada. However, the institutional solutions that have been generated to address these central-subnational tensions vary across cases.

In the United States, Congress holds the authority to sign trade agreements, although negotiating authority can be (and often is) temporarily delegated to the president. In the event that state-level governments violate the international trade commitments made by the federal government, these subnational authorities are explicitly shielded from being sanctioned either directly by international rulings or by the federal government. These protections were established in the 1990s, even as US commitments in the Uruguay Round and in NAFTA extended liberalizing obligations into deeper "behind the border" issues such as government procurement, sanitary and phytosanitary standards, and services liberalization.[39] Authority over many of these issues lay with states, in particular that of government procurement. Rather than extending the reach of the US international commitments down to the state level, the US implementation law for the Uruguay Round agreement explicitly protects subnational governments from WTO sanction by building in "layers of political insulation around the states."[40] This law rules out the ability of foreign actors to sue US subnational authorities in federal court, leaving WTO state-to-state dispute settlement as the main channel for addressing any concerns of unfair trade. In the event a panel rules that a US subnational government has violated WTO rules, the law stipulates that state-level governments are not directly bound by the panel decision. Instead, the United States Trade Representative is then obliged to engage in consultations with the relevant state government to come up with an arrangement that is acceptable to both parties. Even if the federal government then decides to file suit against the state, the US court considers the case anew without regard to the WTO panel's decision, placing the onus on the federal government to make its case against the state.

While the EU is not a federation, it nonetheless operates in quasi-federal ways in the realm of economic policymaking, with supranational authority becoming concentrated in the EC over time. The 1957 Treaty of Rome granted the EC the authority to negotiate trade agreements, and in 1968 the EC gained authority over the enactment of antidumping and countervailing duties. From 1971 onward, the European Court of Justice made a number of decisions that clarified the EC's competency over trade.[41] Within the EC itself, the Directorate-General for Trade gradually grew to become the key agency in charge of trade policy.[42] This centralization of authority over trade policy at the EU level has created a layer of political insulation lowering the probability of sanction at the nation-state level. First, while EU institutions are bound by international trade law, there is greater uncertainty over the degree to which WTO laws apply to individual nations within the EU.[43] Second, much of EU decision making operates by consensus. The fact that trade-related complaints are brought against the EU rather than individual EU states means that a collective action problem emerges whereby the threat of sanction is diffusely distributed across multiple national governments, each of which could decide to resist complying with the WTO. The quasi-federal governing structure of the EU therefore generates lower probabilities of sanction from trade rules for its individual members relative to the EC.

In contrast to the United States and the EU, trade authority in Canada is shared between the federal government and its provinces. While the power to enter into trade agreements is vested in the federal government, it does not enjoy exclusive jurisdiction over foreign commerce on terms comparable to that set out in the US Constitution or EU laws. Rather, the federal and provincial governments both possess the authority to implement international agreements.[44] Additionally, provincial jurisdiction over many "behind the border" issues that are now the focus of trade negotiations (e.g., procurement, energy, and labor, health, and safety standards) means that subnational authority is deeply implicated in Canadian trade agreements. This led, for example, to the direct participation of Canadian provincial governments in trade negotiations with the EU to establish the Canada-EU Free Trade Agreement.[45]

This diffusion of trade authority generates a fair degree of uncertainty over who bears the cost of sanction in a trade dispute. Although the federal government is responsible for Canada's international treaty commitments, there is no law or mechanism specifying how to hold provinces to account for Canada's international obligations, generating broad collective action problems when it comes to implementing trade agreements. In the absence of greater legal clarity, the federal government has had to bear the costs of disputes involving provincial violations of international agreements. For example, in 2009 the US company AbitibiBowater filed a case against Canada under Chapter 11 of NAFTA over the decisions of the Newfoundland and Labrador governments to expropriate the company's assets. The federal government settled the claim and paid CAD$130 million to the company in compensation. None of these costs were recouped from the provinces.[46] Developmental policies enacted by Canadian provinces have also generated trade disputes that are difficult to resolve. For example, Canadian provinces have the authority to set the logging fees charged to forestry companies operating on Crown land. The United States claims that these fees have been set artificially low, amounting in practice to an unfair subsidy to Canadian forestry companies, generating a trade dispute that has been difficult to resolve. While the Canadian government is responsible for managing this matter with the United States, Canada has no jurisdiction over the provincial policy that triggered the dispute.[47]

This dynamic—where the federal government is more likely to bear the costs of a trade agreement dispute relative to subnational levels—plays out also in large decentralized developing countries such as Mexico. For example, in 1997 the US company Metalclad filed a NAFTA Chapter 11 case against Mexico over a municipal government's refusal to issue a construction permit to the company despite higher level approvals. While the NAFTA violation occurred at the municipal level, it was the Mexican federal government that had to deal with the dispute and bear the cost of damages when it lost the case.[48]

This short discussion illustrates that from China to the EU and North America, across a range of regime types and stages of development, we find that large decentralized states face a common problem in dealing with international trade agreements: that of varying exposure to international sanction, with concentrated authority and responsibility at the federal/central level and more diffuse threats of sanction at lower levels. This weaker threat of punishment creates a collective action problem: subnational authorities enjoy substantial scope to enact economic strategies that are misaligned with central policies and might even violate international commitments, but the costs of external sanction end up being borne at higher levels of administration.

The specific reason for this center-local tension, of course, will tend to vary across cases. In developing countries, weaker state capacity to enforce laws is more likely to be a salient factor driving a lower threat of sanction at subnational levels. For China, Section 2 of its accession protocol stipulates that China's WTO obligations are to be administered in a uniform manner, including at subnational levels. According to these commitments, the threat of sanction for local governments should be high. However, the central government's poor monitoring and enforcement capacity means that the de facto probability of sanction faced by provincial, city, and county authorities ends up being much weaker. For developed countries, formal legal structures delineating the rights of subnational authorities may be a more relevant factor in determining the threat of sanction across administrative levels. US state-level governments are explicitly shielded from the threat of international sanction by domestic legislation, while in Canada the lack of legal clarity over implementation responsibilities leaves the federal government bearing the dominant burden of costs arising from international economic disputes. While the application of international law on EU member states is also unclear, the threat of sanction is low because any single member state can block the move to comply with a WTO panel ruling.

Rather than treating WTO entry as a disciplining force that will bring convergence in economic governance, the evidence provided in this book suggests that in the case of large decentralized or quasi-federalized states, the tension between international agreements and subnational economic policies might generate intractable conflicts for national authorities. Importantly, these dynamics seem to cut across regime types and levels of development, applying as much to a single-party authoritarian regime such as China as to a developed democracy such as Canada. Consequently, an important area for future research involves connecting the subnational politics of other large decentralized countries to the politics of global integration.

Authoritarian Responses to Globalization

How, then, does regime type affect the generalizability of this book's arguments? Might other authoritarian regimes face similar domestic political dynamics in their economic integration efforts? It might seem logical to assume that compared to democratic regimes, authoritarian states will tend to have an easier time managing the domestic politics of signing on to trade agreements or entering international organizations such as the WTO. After all, reason might suggest that once an authoritarian government makes the decision to commit to external liberalization, implementation would proceed more smoothly compared to democracies, given top-down political control. However, this study has underscored the point that while the opacity commonly associated with authoritarian regimes might render political conflict less visible, authoritarian governance does not translate into automatic obeisance or a convergence of interests within the state apparatus. Although authoritarian regimes are often associated with repression—of civil society, opposition parties, and media—another feature of authoritarian rule might be that the most significant political competition ends up being almost entirely internalized *within* the state. In developed democracies, disagreements over economic policy frequently manifest themselves in differences between political parties, with policy trajectories shifting depending on which party is in power. The findings of this study suggest that conflict over such issues in nondemocratic regimes might show up instead across competing parts of the state.

Generalizing the case of China's WTO entry to all other authoritarian systems might be difficult. As scholars have pointed out, there are several types of authoritarianism that operate according to different political logics.[49] I propose that the arguments presented in this book specific to political contestation in China might primarily extend to other single-party regimes rather than other authoritarian types. Potential lines of contestation might lie between the party leadership and the state bureaucracy in these single-party systems, between important economic agencies or across competing factions, depending on which parts of the governing apparatus stand to gain or lose from new global economic rules.

One reason to expect similar dynamics at play in other one-party regimes is that authoritarian states arguably have more to lose from WTO entry compared to developed democracies. This is due to the potential for these external rules to affect two dimensions of autocratic interests. First, the economic effects of WTO entry alter the ability of governmental actors to generate economic growth. In the case of China, economic growth forms part of the intense competition between subnational leaders for promotion up bureaucratic ranks. The dynamics found in the Chinese case might therefore generalize to those regimes that—like China—rely to a certain degree on economic growth to produce the performance legitimacy often associated with nondemocratic systems and less so for regimes based more on extraction and predation, or charismatic rule. Second, WTO rules governing state behavior threaten to reshape the distribution of bureaucratic discretion and control over resources. In developed democracies there is a relatively clear distinction between party and state, and politicians compete for resources and influence through legislative and electoral contests. In authoritarian regimes, by contrast, bureaucratic control over resources feeds directly into the ability of state actors (who are also politicians) to engage in political competition. Therefore, to the extent that new trade agreements are increasingly focused on the harmonization of domestic regulations, we might expect external rules to intensify conflict within authoritarian regimes over who retains influence over economic policy.

Another key factor that affected the trajectory of WTO-era economic policy in China is the embeddedness of leadership networks within different parts of the state (the central bureaucracy versus the provinces). While not all authoritarian regimes operate on the same fusion of party and state that is at the core of China's communist system, we might expect similar types of embeddedness to affect policymaking in single-party regimes, or semicompetitive regimes dominated by a single party that systematically draws its members from within the bureaucracy. In such cases, then, party patronage networks would extend to the bureaucracy as well as outside of it, thereby potentially generating the same types of punishment dilemmas as those faced by Chinese leaders.

As the terms of economic globalization steadily reach behind the borders to focus on issues traditionally restricted to the realm of domestic regulation and as such trade agreements increasingly expand to include nondemocratic regimes, a rich area of future research involves exploring the bureaucratic politics of trade liberalization in autocracies. As this study makes clear, globalization has the potential to deepen lines of conflict within single-party regimes, leading to unexpected policy divergence across multiple dimensions and over time.

A

Appendix to Chapter 3

Endogeneity Checks

Pre-WTO Trends

To test if sectors that experienced bigger tariff cuts were already on a different policy trajectory compared to sectors for which tariff reductions were smaller, I regress the topic proportion of the selected topics for industries in the pre–World Trade Organization (WTO) years on the total industry tariff reduction in the WTO years (2002–2007), with the key explanatory variable being an interaction term between the size of the tariff reduction and a linear time trend. If an industry that would experience a high tariff cut at WTO entry was already on a trajectory of, for example, increasing developmental policies, then this would show up as a positive coefficient on the interaction term. Table A.1 shows the coefficient on the interaction term for all three strategies at the central,

Table A.1 Pre-WTO (1998–2001) industry policy trends and WTO tariff cuts

Interaction Term	Regulatory	Developmental	Directive
Pre-WTO years X WTO Tariff Change (Center)	0.0001 (0.001)	0.001 (0.001)	0.00004 (0.0002)
Pre-WTO years X WTO Tariff Change (Province)	0.0001 (0.0004)	0.001 (0.001)	0.0002 (0.0005)
Pre-WTO years X WTO Tariff Change (Local)	0.001 (0.001)	0.0001 (−0.0003)	0.002 (0.001)

*p < 0.05, **p < 0.01, ***p < 0.001

provincial, and local levels. There is no significant correlation between pre-WTO trends and WTO tariff reductions for any of the strategies at any of the administrative levels of government.

Did Domestic Interests Affect Tariff Reductions?

To test if China's WTO tariff reductions are endogenous to domestic industrial or bureaucratic interests, I estimate the relationship between industry characteristics in the pre-WTO period (1998–2001) and tariff changes in the WTO period. If, for example, sectors dominated by state ownership were given protection, we would observe a negative relationship between a sector's state-owned enterprise (SOE) output share and its WTO tariff reduction. The characteristics include

- *Industry exports.* The government might allow larger tariff cuts for exporting industries that are less likely to be threatened by import competition.
- *Industry output.* Industries with a large output size might represent products with a large domestic market, leading to a desire to protect such industries.
- *Industry employment.* To maintain social stability, the government might wish to protect industries that employ a large number of workers.
- *SOE share of industry output.* The government might wish to protect industries in which it has a stronger ownership stake.
- *Foreign and nonmainland share of industry output.* Sectors dominated by foreign and nonmainland enterprises are likely to be more liberalized, potentially leading to greater willingness to allow import competition.

The results are shown in Table A.2. I run two estimations, first regressing the WTO change in tariffs on the average pre-WTO levels of industry characteristics (column 1), then on the change in pre-WTO levels of industry characteristics (column 2). The results show no significant correlations between industry characteristics and WTO tariff reductions.

Regression Tables

Main Estimation Results

Table A.3 shows the results of regressions estimating the relationship between tariff cuts and the adoption of different strategies by the central government. Table A.4 shows the results of regressions estimating the relationship between tariff cuts and the adoption of different strategies by the provincial government. Table A.5 shows the results of regressions estimating the relationship between tariff cuts and the adoption of different strategies by the local government.

Table A.2 Do industry characteristics predict tariff cuts?

	(1)	(2)
Industry exports (RMB bn)	0.016 (0.049)	
Industry output (RMB bn)	0.032 (0.025)	
Industry employment (RMB bn)	1.315 (3.293)	
SOE output share	1.020 (4.37)	
Foreign and nonmainland output share	3.126 (3.648)	
Change in industry exports		0.030 (0.690)
Change in industry output		0.132 (0.183)
Change in industry employment		2.802 (37.949)
Change in SOE output share		−22.087 (28.301)
Change in foreign & nonmainland output share		6.955 (26.234)
Constant	4.004* −1.849	5.807*** (0.701)
Observations	118	118

$*p < 0.05$, $**p < 0.01$, $***p < 0.001$
All covariates are pre-WTO averages (1998–2001).

Supporting Tables and Estimation Results for Causal Mechanism Tests

Table A.6 shows the minimum, mean, and maximum values of the provincial autonomy variable across different years (1998–2007).

The Herfindahl-Hirschman index is a measure commonly used to measure industry fragmentation. For example, if a province's industrial base consisted entirely of one industry, its diversity score would be 0, indicating zero diversity. Conversely, if a province's industrial base consisted of four industries with equal output shares, its diversity score would be higher, indicating greater diversity compared to the earlier example. Table A.7 shows the minimum, mean, and maximum values of the industry diversity variable across different years (1998–2007).

Table A.8 shows the results of a regression estimating the relationship between the change in provincial accountability as trade openness increases (represented

Table A.3 Central government responses

	Central Strategies		
	Regulatory	Developmental	Directive
Tariff (inverse)	0.0030*	−0.0001	0.0002
	(0.0013)	(0.0012)	(0.0004)
Tariff standard deviation	0.0021	0.0022	0.0002
	(0.0017)	(0.0017)	(0.0012)
Industry output (RMB bn)	−0.0001	0.00002	−0.00002
	(0.00004)	(0.00003)	(0.00002)
Industry exports (RMB bn)	−0.00001	−0.00002	0.00004
	(0.0001)	(0.0001)	(0.00003)
Industry employment (millions)	0.0028	−0.0387	−0.0034
	(0.0579)	(0.0595)	(0.0209)
SOE output share	0.0342	0.073	0.0116
	(0.067)	(0.0673)	(0.0421)
Foreign output share	0.046	0.0925	0.024
	(0.1026)	(0.1212)	(0.0674)
Year fixed effects	yes	yes	yes
Industry fixed effects	yes	yes	yes
Observations	659	659	659

$*p<0.05, **p<0.01, ***p<0.001$
Standard errors clustered by industry.

Table A.4 Provincial government responses

	Provincial Strategies		
	Regulatory	Developmental	Directive
Tariff (inverse)	0.0021**	0.0025*	−0.0005
	(0.0007)	(0.001)	(0.001)
Tariff standard deviation	0.0018	0.0004	−0.0003
	(0.0016)	(0.001)	(0.0012)
Industry output (RMB bn)	0.00002	−0.00003	−0.00005
	(0.00002)	(0.00003)	(0.00002)
Industry exports (RMB bn)	0.00003	0.0001	−0.0001
	(0.00005)	(0.0001)	(0.00005)
Industry employment (millions)	−0.0673	−0.0257	0.034
	(0.0474)	(0.033)	(0.0306)
SOE output share	0.0052	−0.0365	−0.0423
	(0.0673)	(0.0724)	(0.0718)
Foreign output share	−0.0554	0.0374	0.0954
	(0.0899)	(0.074)	(0.1049)
Year fixed effects	yes	yes	yes
Industry fixed effects	yes	yes	yes
Observations	531	531	531

$*p<0.05, **p<0.01, ***p<0.001$
Standard errors clustered by industry.

Table A.5 Local government responses

| | Local Strategies | | |
	Regulatory	Developmental	Directive
Tariff (inverse)	−0.0016***	−0.0002	0.0048***
	(0.0004)	(0.0011)	(0.0011)
Tariff standard deviation	−0.0028**	0.0022	0.0021
	(0.001)	(0.0016)	(0.0027)
Industry output (RMB bn)	−0.00001	−0.00003	−0.00005
	(0.00003)	(0.0001)	(0.00005)
Industry exports (RMB bn)	−0.00003	−0.0001	−0.00002
	(0.00004)	(0.0002)	(0.0001)
Industry employment (millions)	−0.0041	0.1149	−0.098
	(0.0287)	(0.0824)	(0.1184)
SOE output share	0.027	0.0037	−0.0217
	(0.0405)	(0.0929)	(0.1482)
Foreign output share	0.0793	0.1691	0.1037
	(0.0734)	(0.1237)	(0.1702)
Year fixed effects	yes	yes	yes
Industry fixed effects	yes	yes	yes
Observations	413	413	413

$*p < 0.05$, $**p < 0.01$, $***p < 0.001$
Standard errors clustered by industry.

Table A.6 Summary statistics for provincial autonomy variable

Year	Min	Mean	Max
1998	0.33	0.72	1.00
1999	0.30	0.69	0.94
2000	0.36	0.68	0.88
2001	0.25	0.65	0.88
2002	0.42	0.64	0.85
2003	0.38	0.64	0.83
2004	0.33	0.61	0.86
2005	0.36	0.61	0.86
2006	0.30	0.59	0.86
2007	0.42	0.59	0.86

Table A.7 Summary statistics for industry diversity variable

Year	Min	Mean	Max
1998	0.8176	0.9508	0.9831
1999	0.8150	0.9472	0.9832
2000	0.8024	0.9381	0.9823
2001	0.7982	0.9358	0.9823
2002	0.7468	0.9318	0.9820
2003	0.7205	0.9316	0.9812
2004	0.7284	0.9245	0.9814
2005	0.7424	0.9248	0.9805
2006	0.7890	0.9294	0.9804
2007	0.7981	0.9255	0.9809

Table A.8 Response of provincial developmental strategy to change in accountability

	Provincial Developmental Strategies
Accountability X Trade openness	−0.001***
	(0.0002)
Accountability	0.0002
	(0.0002)
Trade openness	−0.0004**
	(0.0001)
FDI share of GDP	−0.001
	(0.001)
Industry diversity	0.001
	(0.001)
SCE share of GDP	0.0001
	(0.001)
GDP	0.000
	(0.000)
Log GDP per capita	0.0001
	(0.0002)
Fiscal balance	0.00001
	(0.00001)
Unemployment	−0.00001
	(0.00002)
Year fixed effects	yes
Province fixed effects	yes
Observations	284

*$p < 0.05$, **$p < 0.01$, ***$p < 0.001$
Standard errors clustered by province.

Table A.9 Response of provincial developmental strategy to change in industry diversity

	Provincial Developmental Strategies
Industry diversity X WTO exposure	0.029* (0.012)
Industry diversity	0.004* (0.002)
WTO exposure	−0.024** (0.009)
Accountability	0.0001 (0.0001)
FDI share of GDP	0.0003 (0.001)
SCE share of GDP	0.001* (0.0004)
GDP	0.000 (0.000)
Log GDP per capita	0.0002 (0.0003)
Fiscal balance	0.00001 (0.00001)
Unemployment	0.00002 (0.00002)
Year fixed effects	yes
Province fixed effects	yes
Observations	284

*p < 0.05, **p < 0.01, ***p < 0.001
Standard errors clustered by province.

by the interaction term "*Accountability X Trade openness*") and the provincial adoption of developmental strategies.

Table A.9 shows the results of a regression estimating the relationship between the change in provincial industry diversity as WTO exposure increases (represented by the interaction term "*Industry diversity X WTO exposure*") and the provincial adoption of developmental strategies.

B

Appendix to Chapter 4

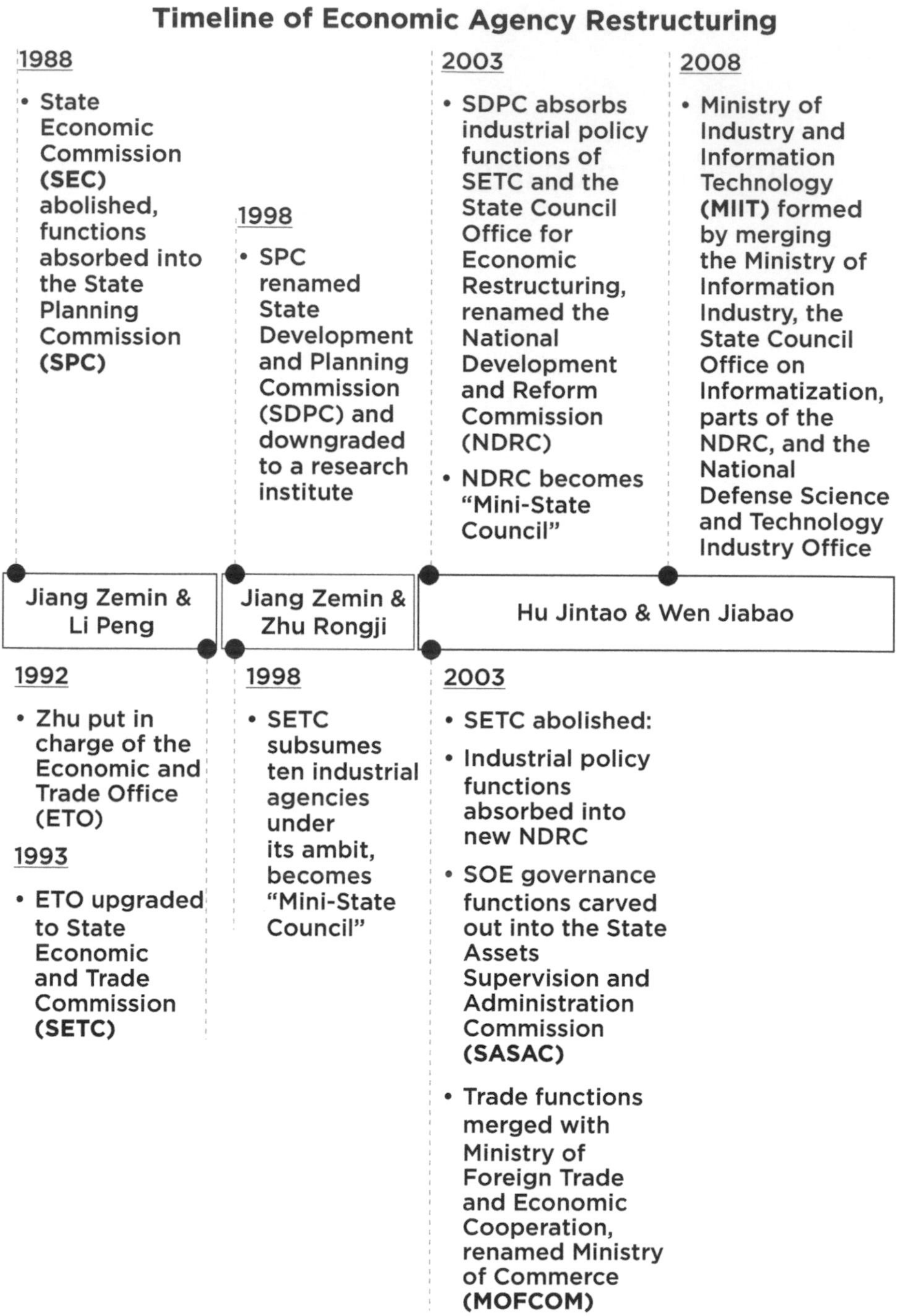

Timeline of Economic Agency Restructuring

1988

- State Economic Commission **(SEC)** abolished, functions absorbed into the State Planning Commission **(SPC)**

1998

- SPC renamed State Development and Planning Commission (SDPC) and downgraded to a research institute

2003

- SDPC absorbs industrial policy functions of SETC and the State Council Office for Economic Restructuring, renamed the National Development and Reform Commission (NDRC)
- NDRC becomes "Mini-State Council"

2008

- Ministry of Industry and Information Technology **(MIIT)** formed by merging the Ministry of Information Industry, the State Council Office on Informatization, parts of the NDRC, and the National Defense Science and Technology Industry Office

| Jiang Zemin & Li Peng | Jiang Zemin & Zhu Rongji | Hu Jintao & Wen Jiabao |

1992

- Zhu put in charge of the Economic and Trade Office (ETO)

1993

- ETO upgraded to State Economic and Trade Commission **(SETC)**

1998

- SETC subsumes ten industrial agencies under its ambit, becomes "Mini-State Council"

2003

- SETC abolished:
- Industrial policy functions absorbed into new NDRC
- SOE governance functions carved out into the State Assets Supervision and Administration Commission **(SASAC)**
- Trade functions merged with Ministry of Foreign Trade and Economic Cooperation, renamed Ministry of Commerce **(MOFCOM)**

Figure B.1. Timeline of economic agency restructuring

List of Interviews

The identities of interview subjects have been kept anonymous (except in cases where the interviewee explicitly agreed to be named) in accordance with the research protocol approved by Harvard University's Committee on the Use of Human Subjects. The list below provides a subject ID for each interviewee as well as a general description and the location, month and year of each interview.

Subject ID	Description	Location	Date
B01	Academic	Beijing	June 2012
B02	Academic	Beijing	July 2012
B03	Academic	Beijing	June 2013
B04	Government researcher	Beijing	June 2013
B05	Journalist	Beijing	July 2013
B06	Government researcher	Beijing	July 2013
B07	Academic	Beijing	July 2013
B08	Academic	Beijing	June 2014
B09	Government researcher	Beijing	July 2014
B10	Retired academic	Beijing	October 2014
B11	Academic	Beijing	October 2014
B12	Academic	Beijing	October 2014
B13	Lawyer	Beijing	October 2014
B14	Academic	Beijing	November 2014
B15	Central government official	Beijing	November 2014
B16	Foreign businessperson	Beijing	November 2014
B17	Central government official	Beijing	November 2014
B18	Former government official	Beijing	March 2015
B19	Government researcher	Beijing	March 2015

(continued)

Subject ID	Description	Location	Date
B20	Government researcher	Beijing	March 2015
B21	Central government official	Beijing	March 2015
B22	Central government official	Beijing	April 2015
B23	Academic	Beijing	April 2015
B24	Foreign businessperson	Beijing	May 2015
B25	Chinese think tank member	Beijing	May 2015
B26	Academic	Beijing	May 2015
B27	Academic	Beijing	May 2015
B28	Foreign businessperson	Beijing	June 2015
B29	SOE manager	Beijing	June 2015
B30	Foreign businessperson	Beijing	July 2015
B31	Private businessperson	Beijing	July 2015
B32	Government researcher	Beijing	June 2016
B33	Private businessperson	Beijing	June 2016
B34	Foreign businessperson	Beijing	June 2016
B35	Government association official	Beijing	July 2016
B36	Government association official	Beijing	July 2016
S01	Government researcher	Shanghai	June 2013
S02	Government researcher	Shanghai	August 2014
S03	Government researcher	Shanghai	April 2015
S04	Private businessperson	Shanghai	April 2015
S05	Private consultant	Shanghai	April 2015
S06	Government researcher	Shanghai	April 2015
S07	Government researcher	Shanghai	April 2015
S08	Academic	Shanghai	April 2015
S09	Graduate student	Shanghai	April 2015
S10	Private businessperson	Shanghai	April 2015
S11	Private businessperson	Shanghai	April 2015
S12	Private businessperson	Shanghai	April 2015
SZ01	Government-affiliated center	Suzhou	April 2015
SZ02	Foreign businessperson	Suzhou	April 2015
GZ01	Academic	Guangzhou	November 2014
GZ02	Foreign businessperson	Guangzhou	November 2014
GZ03	Lawyer	Guangzhou	November 2014
GZ04	Foreign businessperson	Guangzhou	November 2014
Y01	SOE manager	Yunnan	December 2014
Y02	Provincial government official	Yunnan	December 2014
Y03	Academic	Yunnan	December 2014
Y04	Private businessperson	Yunnan	December 2014
Y05	Academic	Yunnan	December 2014
Y06	Academic	Yunnan	December 2014

Subject ID	Description	Location	Date
Y07	SOE manager	Yunnan	December 2014
Y08	SOE manager	Yunnan	December 2014
Y09	SOE manager	Yunnan	December 2014
Y10	SOE manager	Yunnan	December 2014
G01	International organization official	Geneva	February 2015
G02	International organization official	Geneva	February 2015
G03	International organization official	Geneva	February 2015
P01	Academic	Paris	February 2015
SG01	Foreign businessperson	Singapore	June 2015
C01	International organization official	Cambridge, MA	September 2013
C02	Government researcher	Cambridge, MA	October 2015
C03	Lawrence Summers	Cambridge, MA	March 2016
T01	Academic	Skype	July 2013
T02	Academic	Telephone	November 2014
T03	Academic	Telephone	April 2015
T04	Journalist	Telephone	June 2015
T05	Foreign businessperson	Telephone	June 2015
T06	Jeffrey Bader	Telephone	April 2016
T07	Former US official	Telephone	April 2016

Notes

Introduction

1. United Nations Statistical Division, "UN Comtrade Database," https://comtrade.un.org.

2. The White House, "Joint Press Conference of the President and Premier Zhu Rongji of the People's Republic of China," April 8, 1999, https://clintonwhitehouse4.archives.gov/WH/New/html/19990408-1109.html.

3. William Clinton, "Remarks at the Paul H. Nitze School of Advanced International Studies," The American Presidency Project, March 8, 2000, https://www.presidency.ucsb.edu/documents/remarks-the-paul-h-nitze-school-advanced-international-studies.

4. Interview with former assistant USTR Jeffrey Bader, 2016.

5. World Trade Organization, "Protocol on the Accession of the People's Republic of China," November 23, 2001.

6. "Commentary Hails China's Accession to WTO," *Xinhua News,* December 11, 2001, retrieved June 16, 2018, from Factiva Database.

7. "Seize Opportunity, Meet Challenge, and Participate in Economic Globalization," *People's Daily,* December 19, 2001, http://en.people.cn/200112/19/eng20011219_87044.shtml.

8. "Clean-up of WTO-Contradictory Laws a Priority—Senior Official," *Xinhua News,* December 12, 2001, http://www.china.org.cn/english/government/23542.htm; "China to Build Impartial, Efficient Judicial System—Li Peng," *Xinhua News,* December 12, 2001, http://german.china.org.cn/english/government/23546.htm; "People's Daily Commentary on SOE Reform," *Xinhua News,* December 11, 2001, retrieved June 16, 2018, from Factiva Database.

9. Erik Eckholm, "One Giant Step for Mr. Jiang's China," *New York Times,* November 21, 1999.

10. United States Trade Representative, "2018 Report to Congress on China's WTO Compliance," February 2019, 5.

11. Ibid., 6.

12. Ibid., 51.

13. Campbell and Ratner 2018.

14. Wang et al. 2018.

15. Ana Swanson, "U.S. Joins Europe in Fighting China's Future in WTO," *New York Times,* November 29, 2017.

16. United States Trade Representative, "Following President Trump's Section 301 Decisions, USTR Launches New WTO Challenge against China," March 23, 2018, https://ustr.gov/about-us/policy-offices/press-office/press-releases/2018/march/following-president-trump%E2%80%99s-section; United States Trade Representative, "President Trump Announces

181

Strong Actions to Address China's Unfair Trade," March 22, 2018, https://ustr.gov/about-us/policy-offices/press-office/press-releases/2018/march/president-trump-announces-strong; United States Trade Representative, "Under Section 301 Action, USTR Releases Proposed Tariff List on Chinese Products," April 3, 2018, https://ustr.gov/about-us/policy-offices/press-office/press-releases/2018/april/under-section-301-action-ustr; Ana Swanson, Keith Bradsher, and Katie Rogers, "Trump Threatens Tariffs on $200 Billion in China Goods, Escalating Fight," *New York Times,* June 18, 2018.

17. "What's in the U.S.-China Phase 1 Trade Deal," *Reuters,* January 15, 2020.

18. See, for example, the Ministry of Commerce's statement on China's abidance with WTO rules in "China Follows WTO Rules," *China Daily,* March 24, 2017, http://www.chinadaily.com.cn/business/2017-03/24/content_28662559.htm. Another notable instance was at the 2017 Davos World Economic Forum meeting, where President Xi Jinping's speech in support of free trade and globalization was widely portrayed as a much-needed pushback against growing protectionist sentiments in other parts of the world.

19. The earlier stance of maintaining a low profile is commonly encapsulated in Deng Xiaoping's phrase "taoguang yanghui," loosely translated as meaning "to hide one's capacity and to bide one's time."

20. Ministry of Foreign Affairs of the People's Republic of China, "Xi Jinping Calls for the Building of New Type of International Relations with Win-Win Cooperation at the Core in a Speech at the Moscow State Institute of International Relations," March 23, 2013, https://www.fmprc.gov.cn/mfa_eng/topics_665678/xjpcf1_665694/t1024781.shtml; "Full Text from President Xi Jinping's Speech at the National Committee on United States–China Relations," September 22, 2015, https://www.ncuscr.org/content/full-text-president-xi-jinpings-speech.

21. "Live Broadcast Celebrating the 95th Anniversary of the Founding of the Chinese Communist Party" ("Qingzhu Zhongguo Gongchandang Chengli 95 Zhounian Dahui Zhibo"), *Xinhua News,* July 1, 2016, http://www.xinhuanet.com/politics/zbzg952016/wzsl.htm (my emphasis).

22. "Secure a Decisive Victory in Building a Moderately Prosperous Society in All Respects and Strive for the Great Success of Socialism with Chinese Characteristics for a New Era," *Xinhua News,* November 3, 2017, http://news.xinhuanet.com/english/special/2017-11/03/c_136725942.htm (my emphasis).

23. Ikenberry 2009, 72.

24. See Fewsmith 2001a. For a discussion of how China's leaders used WTO entry as commitment for domestic reform, see Pearson 2001.

25. Wang 2000.

26. Naughton 2007a.

27. Wu 2013.

28. Liaoning Provincial Government, "Circular regarding Liaoning Province 2012–2015 Work Plan for the Rectification and Closure of Metal and Non-metal Mines" ("Guanyu Yinfa Liaoning Sheng Jinshu Fei Jinshu Kuangshan 2012–2015 Nian Zhengdun Guanbi Gongzuo Fang'an de Tongzhi"), 2012, www.pkulaw.cn.

29. Hechi City People's Government, "Circular regarding Hechi City's Implementation Plan for Promoting Indigenous Innovation in Non-ferrous Metal Deep Processing Enterprises" ("Guanyu Yinfa Hechi Shi Tuijin Youse Jinshu Shen Jiagong Qiye Zizhu Chuangxin Shishi Fang'an de Tongzhi"), 2008, www.pkulaw.cn.

30. Shanghai City Quality and Technical Supervision Bureau, "Circular regarding the Shanghai Metal Materials Quality Supervision and Inspection Station Passing its Expansion Review" ("Guanyu Shanghai Shi Jinshu Cailiao Zhiliang Jiandu Jianyanzhan Tongguo Kuoxiang Pingshen de Tongzhi"), 2008, www.pkulaw.cn.

31. Zhang and Li 2014.

32. Landry 2008.

33. Landry, Lü, and Duan 2018; Shih, Adolph, and Liu 2012; Zheng 2004.

34. Pearson 2005; Yang 2004.

35. Naughton 2011; Heilmann and Shih 2013.

36. Nathan 2003.

37. Blei and Lafferty 2007.

1. China's WTO Challenge

1. Barton 2006, 45.
2. Ikenberry 2009.
3. Barton 2006, 47.
4. Ibid., 129.
5. A set of flexibilities and exceptions are granted to members with developing country status. Under the Generalized System of Preferences, developed countries can offer concessions such as zero or very low tariffs to imports from developing countries. Developing countries are also given longer implementation periods for acceding to various agreements as well as assistance to strengthen their capacity to engage in WTO-related work.
6. Goldstein 2012, 79.
7. Shaffer 2015.
8. Hoekman and Mavroidis 2000, 71.
9. These information requirements are entirely different from GATT practices in the 1940s, which allowed all correspondences between trade delegates to be privileged and confidential and to stay confidential in perpetuity if a delegate did not formally rescind the confidentiality request within three years (Goldstein and Martin 2000).
10. Pelc 2011.
11. "WTO Successfully Concludes Negotiations on China's WTO Entry," World Trade Organization Press Release, September 17, 2001, https://www.wto.org/english/news_e/pres01_e/pr243_e.htm.
12. Feng 2006, 201.
13. Interview with former assistant USTR Jeffrey Bader, 2016.
14. Pearson 2001, 342.
15. These include Pearson 2001; Fewsmith 2000; Feng 2006.
16. Interview with Bader, 2016.
17. Interview with former US treasury secretary Larry Summers, 2016.
18. "I want them to understand two things," Barshefsky said. "First, we would like them in the WTO, but, second, the price is not cheap." Quoted in Elsa Walsh, "The Negotiator," *New Yorker*, March 18, 1996, 86.
19. Devereaux, Lawrence, and Watkins 2006, 263.
20. US business interests were represented by the US Chamber of Commerce, with major players coming from a broad swath of industries, including AIG, Boeing, Motorola, General Motors, Walmart, General Electric, and farm groups (Interview with Bader, 2016).
21. Interview with Bader, 2016.
22. Interview with former US official, 2016.
23. Interview with Bader, 2016.
24. William Clinton, "Full Text of Clinton's Speech on China Trade Bill," *New York Times*, March 9, 2000.
25. Lawrence Summers, "Treasury Secretary Lawrence H. Summers Testimony before the House Committee on Ways and Means," US Department of the Treasury, May 3, 2000, https://www.treasury.gov/press-center/press-releases/Pages/ls601.aspx.
26. Pearson 2001.
27. Fewsmith 2000.
28. Pearson 2001, 364.
29. Wang 2002.
30. Pearson 2001, 361.
31. Willy Lam, "Open Door May be Closed," *South China Morning Post*, November 26, 1997.
32. Zhu 2011a, 359.
33. The Twenty-One Demands were territorial and governmental concessions that Japan had demanded of China in World War I.
34. Zong 2002b, 49.
35. Willy Lam, "Zhu Deal Backlash," *South China Morning Post*, May 5, 1999.
36. Pearson 2001; Fewsmith 2000.
37. Willy Lam, "Beijing Rift over Accession," *South China Morning Post*, May 26, 1999.

38. Mary Kwang, "Zhu's Market-Opening Offers 'Not Concessions,'" *Straits Times*, May 8, 1999.

39. Pearson 2001, 345.

40. Devereaux, Lawrence, and Watkins 2006.

41. Interview with Bader, 2016.

42. Sticking points included foreign equity shares on telecommunications and insurance, automobile financing, and market access and equity shares for audiovisuals, among others. A Chinese transcript of Zhu's meetings with the US delegation on November 13 and 15 can be found in Zhu 2011a.

43. Interview with Bader, 2016.

44. Interview with former US official, 2016.

45. Devereaux, Lawrence, and Watkins 2006, 276–77.

46. Author's translation (Zhu 2011a, 367).

47. In Zhu's words, he had to stand firm on the issue in order to be able to account to the Chinese people for the deal (Zhu 2011a, 358).

48. Zhu 2011a, 361.

49. Interview with former US official, 2016.

50. Joseph Kahn, "Details of China Trade Pact Released by U.S.," *New York Times*, March 15, 2000.

51. Daniel Rosen, "China and the World Trade Organization: An Economic Balance Sheet," Policy brief 99-06, Peterson Institute for International Economics, 1999.

52. Qin 2007. This assessment does not imply that WTO entry did not bring substantial benefits to China or that the event did not bring dislocations to the US economy. One study found that while imports from China from 1995 to 2011 accounted for 2 million job losses, the growth in US exports over the same period of time meant that on net, demand for jobs in the United States grew by about 1.7 million (Feenstra and Sasahara 2018). At the same time, China's WTO entry, while made on stringent terms, undoubtedly brought substantial economic benefits. The Chinese Ministry of Commerce's own estimates show that Chinese trade in goods expanded eightfold between 2001 and 2014, making China the world's largest trading nation. China's GDP also grew from US$1.3 trillion to US$10 trillion during that same period, behind only the United States in size. See Yuan, "Looking Back 14 Years after Accession: Case of China," Third China Roundtable on WTO Accessions, World Trade Organization, June 2, 2015, https://www.wto.org/english/thewto_e/acc_e/Session2YuanYuanPostAccessionLooking back14yearafter.pdf.

53. Interview with former US official, 2016. In contrast, in the United States international treaties require domestic legislation or implementing regulations in order to have legal force unless these treaties are considered self-executing. See Michael Garcia, "International Law and Agreements: Their Effect upon U.S. Law," Congressional Research Service Report No. RL32528, 2015.

54. Bound tariffs are the tax rates that WTO members have formally agreed to, while applied tariffs are the tax rates that are actually in use. In many developing countries, there is a large gap (or "water") between bound and applied tariffs, giving these countries the ability to raise their applied tariffs without violating any WTO rules.

55. Qin 2003.

56. A number of scholars have examined these aspects of China's protocol of accession. This section draws primarily on detailed analyses by Qin 2003 and Gao 2007.

57. National treatment requires providing treatment to others equal to that accorded to one's nationals.

58. In other words, WTO members do not have to use Chinese domestic prices to establish if there is dumping behavior but can use the prices of a surrogate country. This gives WTO members substantial discretion in choosing benchmark prices in their dumping cases.

59. Under WTO rules, members can retaliate immediately.

60. It should be noted that under WTO rules, developing country status is self-designated rather than accorded based on any set of socioeconomic standards. Hence, China agreed to self-designate that it did not qualify for the flexibilities accorded to other developing countries in the WTO.

61. Interview with Bader, 2016.
62. Author's translation, drawn from Li 2001, 4.
63. Author's translation, drawn from Liang 2001, 7.
64. Literature on these topics is extensive. Examples include Rosendorff and Milner 2001 on institution formation, Davis 2012 and Pelc 2014 on legal adjudication, and Goldstein, Rivers, and Tomz 2007 and Davis and Wilf 2011 on membership and accession.
65. Gourevitch 1978; Ruggie 1982; Goldstein and Martin 2000.
66. Krasner and Weinstein 2014, 136.
67. Oatley 2011.
68. Rogowski 1987a; Frieden 1991; Hiscox 2001.
69. Grossman and Helpman 1994.
70. These mechanisms include audience costs and sinking costs, reputation, and delegation as well as domestic politics channels such as information, uncertainty, and signaling.
71. Harpaz 2011; Zhang and Li 2014.
72. Webster 2014; Oh 2015.
73. Chang 2002; Wade 2003; Rodrik 2011.
74. Amsden and Hikino 2000; Weiss 2005; Shadlen 2005.
75. Clift and Woll 2012, 311.
76. Additionally, some highly developed cities are accorded deputy provincial status between that of the province and the prefecture.
77. Lieberthal and Oksenberg 1988; Bachman 1992; Chung 2016.
78. Baum and Shevchenko 1999; Hurst 2009; Rithmire 2014.
79. Saich 2002. For literature on fragmented authoritarianism, see Lieberthal and Oksenberg 1988; Mertha 2009.
80. Yang 2002; Fewsmith 2001a.
81. DeWoskin 2001; Harwit 2001.
82. Gao 2011; Kennedy and Cheng 2012; Li 2012.
83. Exceptions include Mertha 2005b; Kennedy 2009; Hsueh 2011; Steinfeld 2012; Zeng and Liang 2013.
84. Yang 2004; Qin 2007; Mertha and Zeng 2005.
85. Heilmann, Shih, and Hofem 2013; Naughton 2007b.
86. Hsueh 2011.
87. Pearson 2015.
88. Lieberthal and Oksenberg 1988.
89. The periodization in this section draws on prominent Chinese economist and government adviser Wu Jinglian's September 2008 speech "Institutional Reflections from China's 30 Years of Economic Reform" (Wu 2008).
90. Naughton 2007a, 380–81.
91. Gewirtz 2017; Wu 2013, 40–41.
92. This period is most representative of the famous phrase describing China's reform process as one of "crossing the river by feeling for the stones," often attributed to Deng but actually said by party elder Chen Yun.
93. Weingast, Qian, and Montinola 1995.
94. Qian and Wu 2003.
95. Central Committee of the Chinese Communist Party, "Zhonggong Zhongyang Guanyu Jingji Tizhi Gaige de Jueding" (Decision on Economic System Reform), October 20, 1984, http://cpc.people.com.cn/GB/64.
96. See Chinese economist Xue Muqiao's discussion of this in Xue 2010.
97. Cornick 2014.
98. State Council Development Research Center, "Woguo Chanye Zhengce de Chubu Yanjiu" (Preliminary Study of China's Industrial Policies), March 1987.
99. Zhang 1997.
100. State Council of the People's Republic of China, "Guowuyuan Guanyu Dangqian Chanye Zhengce Yaodian de Jueding" (Decision on Current Industrial Policy Priorities), State Council No. 5, April 10, 1989.
101. In Chinese, the phrase reads "计划于市场有机地结合起来."

102. Fewsmith 2001b.

103. Central Committee of the Chinese Communist Party, "Guanyu Jianli Shehui Zhuyi Shichang Jingji Tizhi Ruogan Wenti de Jueding" (Decision on Several Questions relating to the Establishment of the Socialist Market Economy), November 14, 1993, http://cpc.people.com.cn/GB/64162/134902/8092314.html.

104. Qian and Wu 2003.

105. Wu 2013, 54, 42.

106. Huan 1996.

107. Chen 1996.

108. State Council of the People's Republic of China, "Guowuyuan Jiushi Niandai Guojia Chanye Zhengce Gangyao" (Outline of State Industrial Policy for the 1990s), March 25, 1994.

109. My italics. In Chinese, the phrase on the role of the market reads "充分发挥市场在国家宏观调控下对资源配置的基础性作用."

110. Qian and Wu 2003.

111. Central Committee of the Chinese Communist Party, "Zhonggong Zhongyang Guanyu Guoyou Qiye Gaige he Fazhan Ruogan Zhongda Wenti" (Decision on Major Issues Concerning the Reform and Development of State-owned Enterprises), September 22, 1999.

112. "Jiang Zemin's Speech at the Meeting Celebrating the 80th Anniversary of the Founding of the Communist Party of China," July 1, 2001, China.org, http://www.china.org.cn/e-speech/a.htm. In his speech, Jiang declared that "the Party must always represent the requirements of the development of China's *advanced productive forces*, the orientation of the development of China's *advanced culture*, and the fundamental interests of the *overwhelming majority of the people in China*" (my emphasis). The expansion of the Party to include representation of "advanced productive forces" indicated an inclusion of private entrepreneurs that, Jiang argued, represented a new component of the working class that had emerged through the past decades of reform (Shambaugh 2008, 111–13).

113. Qian and Wu 2003.

114. World Bank 2020.

115. Naughton 2007a, 384–86.

2. A Theory of State Strategies under Global Rules

1. There is extensive literature on the topic of central-local relations in China. For recent overviews, see Chung 2016 and Donaldson 2016.

2. Shirk 1993, 83.

3. Zhang Yi, "Membership of CPC tops 90 Million," *China Daily*, July 1, 2019. The conflicted authority relations are often referred to as the tension between *tiao* (vertical supervision) and *kuai* (local government autonomy). Subnational agencies are charged with implementing policies issued by their central superiors. However, they are also accountable to the leadership in their particular province or locality and are reliant on these leaders for resources and career advancement (Lieberthal and Oksenberg 1988).

4. Norris 2016.

5. Mertha and Zeng 2005, 320.

6. Perry 2011; Rooij 2009.

7. Edin 2003, 35.

8. While the number of noneconomic targets in the cadre evaluation system has expanded over the years, this has had little effect on the primacy of economic growth in shaping the behavior of government leaders. This is because "growth promotion is still listed as the first item on the guidelines and constitutes the largest share of total scores, but also because economic outputs, compared to other soft targets like environmental protection, are more measurable" (Ang 2016, 122–23).

9. Li and Zhou 2005; Landry 2008; Yao and Zhang 2015. I focus primarily on the influence of the evaluation system on local leader behavior rather than the rank-and-file cadres, because the latter tend to stay in the same locality and are relatively less careerist (Ang 2016, 108). Recent studies have suggested that the link between economic growth and promotion is

tightest at the county level and weakens at higher levels of administration (Landry, Lü, and Duan 2018). However, this finding does not capture a broader link between economic growth and political advancement that applies to subnational leaders at all levels. That is, "thriving economies bring numerous personal benefits to local leaders, including opportunities to exert power, command prestige, distribute patronage, and collect personal rents" (Ang 2016, 123).

10. Carpenter 2001.

11. Shih 2008, 54.

12. Scholars have also pointed to the spread of corruption within the CCP, suggesting that leaders might be motivated more by rent seeking than by promotion (Pei 2016). However, the two factors (promotion and rent seeking) are mutually reinforcing, since promotion to a higher level of government is likely to enlarge that leader's opportunities for rent seeking. Therefore, the corruption incentive fuels rather than mitigates the desire for political advancement, and both are positively linked to economic expansion.

13. This study focuses on the impact on WTO entry on governance over manufacturing sectors and does not delve into issues such as services liberalization and intellectual property protection. The focus on manufacturing industrial policies is motivated by the dominance of this sector in China's economy and China's vital role as "workshop of the world." For an excellent study of how external pressure has affected China's intellectual property rights regime, see Mertha 2005b.

14. Grossman and Helpman 1994.

15. See, for example, Mansfield, Milner, and Rosendorff (2000), who note that authoritarian regime preferences on trade liberalization are unpredictable, as they are dependent on leadership preferences.

16. Zhu 2011; Yang and Zhao 2014.

17. While trade liberalization can generally be expected to produce a onetime level increase in output, the empirical relationship between trade and growth rates is much more difficult to establish, as has been widely discussed in the economics literature.

18. Lardy 2014, 45.

19. Shaffer 2015.

20. Lau, Qian, and Roland 2000; Rawski 1999.

21. Leading Small Groups are informal bodies formed to oversee specific tasks or policy priorities that often report directly to the Politburo or its Standing Committee or the State Council, giving them status and influence higher than that of ordinary ministries.

22. Mahoney and Thelen 2010.

23. The relationship between the state and the market has been typologized in many different forms. Evans (1995), for example, suggests four roles that the state can play vis-à-vis the market: the "custodian" (state as regulator), the "demiurge" (state as producer), the "midwife" (state as assistant to supporting new types of entrepreneurial groups or production), and a "husbandry" role (state provision of support to help firms meet global challenges) (13–14). For other examples, see Johnson 1982; Vogel 2018.

24. Johnson (1982) uses the term "market-conforming" to describe Japan's developmental state, described as "a government-business relationship that both enabled the government to achieve genuine industrial policy and also preserved competition and private enterprise in the business world" (29). I use the term "market-shaping" to highlight that developmental measures have the effect of influencing the decisions of economic actors and in fact are often explicitly designed to do so—a feature of the developmental strategy that is overlooked by the term "market-conforming."

25. "Understanding the Sanitary and Phytosanitary Measures Agreement," World Trade Organization, May 1998, http://www.wto.org/english/tratop_e/sps_e/spsund_e.htm.

26. China's Protocol of Accession further specifies regulatory commitments such as the impartial application of rules (Section 2A), transparency (Section 2C), independent judicial review (Section 2D), nondiscrimination of foreign entities (Section 3), and regulatory independence in services (Annex 1A Section V(d)).

27. Zysman 1977; Levy 2006.

28. Rodrik 2011.

29. Evans 1995.
30. Rodrik 2011; Hausmann, Rodrik, and Sabel 2008.
31. For overviews, see Salazar-Xirinachs, Nübler, and Kosul-Wright 2014; Stiglitz, Lin, and Monga 2013.
32. I refer to "state strategies" and "strategic responses" interchangeably—that is, a strategic response is a choice of state strategy.
33. Rodrik 1998; Garrett 2001.
34. See, for example, the public debate between influential Peking University economists Justin Lin and Zhang Weiying over the efficacy of industrial policy in promoting growth. Described in Frank Tang, "To Embrace Market or State: Chinese Economists Debate the Country's Future," *South China Morning Post,* November 12, 2016.
35. See Annex 3 of the Marrakesh Agreement.
36. Ministry of Economy, Trade and Industry of the Government of Japan, "Chapter 18: Monitoring Trade Policies/Measures," in *2015 Report on Compliance by Major Trading Partners with Trade Agreements—WTO, FTA/EPA and IIA,* May 2015, www.meti.go.jp/english/report/data/2015WTO/02_18.pdf.
37. This review process focused specifically on China's WTO compliance, unlike the regular trade policy review mechanism whose scope is much broader and not compliance-specific.
38. Tallberg 2002.
39. Zhang and Li 2014.
40. In Chinese the phrase is 下管一级.
41. Mei and Pearson 2014.
42. Ang 2012.
43. While the state is enmeshed within the party in the Chinese system, I distinguish between the two by focusing on the Politburo Standing Committee as the party leadership and on central government ministries as the bureaucracy.
44. See, for example, the series of papers organized around the theme of the China shock, which examine the expansion of Chinese exports after WTO entry and its impact on domestic politics in the United States (Autor, Dorn, and Hanson 2016).
45. Adhikari and Yang 2002.
46. Yao and Zhang 2015.
47. I assume that each agency's approach to economic governance tends to be largely fixed over time, such that different agencies tend to be oriented toward particular state strategies when advancing their policy agenda.
48. Shaffer 2015.

3. The Hierarchical Politics of WTO Entry

1. "Annual Total Population Data by Province," National Bureau of Statistics of China, http://data.stats.gov.cn/english/easyquery.htm?cn=E0103.
2. Jaros and Tan 2020.
3. Mertha and Zeng 2005. This dynamic is captured by a common phrase in Chinese: *shang you zhengce, xia you duice* (the higher level has measures, but the lower level has countermeasures).
4. WTO liberalization potentially creates export opportunities to the extent that imported inputs are now cheaper, thereby improving the competitiveness of some Chinese goods on world markets.
5. This trend is borne out in data comparing provincial and prefecture industrial composition, with the former persistently found to have greater industry diversity than the latter. For more details, see Tan 2020.
6. This argument is similar to research on how constituency size in the United States affects politicians' trade orientations. Politicians overseeing smaller constituencies tend to be more protectionist, as these constituencies are likely to be part of a more homogenous local economy that is more vulnerable to import competition (Rogowski 1987b; Schiller 1999).

7. Some examples of works that examine internal governance debates in China include Wu 2013; Fewsmith 2001b; Ma 2015.

8. My emphasis.

9. Jing Fu, "Colossal SOEs to Advance WTO Role," *China Daily*, December 8, 2001.

10. My emphasis.

11. The seven core strategic sectors were armaments, power generation and distribution, oil and petrochemicals, telecommunications, coal, aviation and shipping, while the six other sectors were machinery, automobiles, construction, iron and steel, nonferrous metals, and information technology. "China Names Key Industries for Absolute State Control," *China Daily*, December 19, 2006.

12. "CPC Acknowledges Market's Decisive Role," *Xinhua News*, July 26, 2016 (my emphasis).

13. Thun 2006; Segal 2010; Huang 2002; Hsueh 2011.

14. For the repository, see "Chinalawinfo Database," Peking University, www.pkulaw.cn.

15. Blei and Lafferty 2007.

16. For technical details of the textual analysis, including the steps taken to preprocess the texts, the criteria used to select the number of topics, and the process of independently validating the topic selection, see the appendix to Tan 2020.

17. Kennedy and Johnson 2016, 19.

18. Wang 2002.

19. This data on industry characteristics was generated from a National Bureau of Statistics firm-level data set covering all manufacturing sectors from 1998 to 2007 of firms with sales exceeding RMB5 million.

20. Full analysis results are in appendix A.

21. OECD 2009, 173.

22. State Administration of Quality Supervision, Inspection and Quarantine, "Management Measures for the Adoption of International Standards," Order No. 10, 2001, http://www.gov.cn/gongbao/content/2002/cont (accessed April 8, 2020).

23. Anhui Provincial Government, "10th Five-Year Automobile Industry Development Plan" (*Anhui Sheng "Shi Wu" Qiche Chanye Fazhan Guihua*), No. 115 2001, Chinalawinfo Database, http://www.pkulaw.cn.

24. Shandong Provincial Government, "Opinion of the Shandong Provincial People's Government on Accelerating the Development of the Shipbuilding Industry," (*Shandong Sheng Renmin Zhengfu Guanyu Jiakuai Chuanbo Gongye Fazhan de Yijian*), No. 71 2002, Chinalawinfo Database, http://www.pkulaw.cn.

25. Yanbian Korean Autonomous Prefecture People's Government, "Circular of the People's Government of Yanbian Korean Autonomous Prefecture on the Rectification of the Prefecture-wide Small Cement Plant" (*Yanbian Chaoxianzu Zizhi Zhou Renmin Zhengfu Guanyu Qingli Zhengdun Quanzhou Xiao Shuini Chang (Fenmo Zhan) de Tongzhi*), No. 25 2003, Chinalawinfo Database, http://www.pkulaw.cn.

26. Xi'an Municipal People's Government, "Circular on Phasing Out Small-Scale Papermaking Enterprises" (*Xi'an Shi Renmin Zhengfu Guanyu Zhubu Taotai Xiaoguimo Zaozhi Qiye de Tongzhi*), No. 146 2003, Chinalawinfo Database, http://www.pkulaw.cn.

27. Suttmeier, Yao, and Tan 2009.

28. I am grateful to Kyle Jaros and David Bulman for sharing this data.

29. All estimations include province and year fixed effects, with standard errors clustered by province. Province-level data is drawn from the CEIC database, and all data range from 1998 to 2007.

4. Explaining the Rise of State Capitalism

1. Naughton and Tsai 2015; Heilmann and Shih 2013; Wu 2016.

2. One reason this trend was not picked up by the previous analysis is that chapter 3 focused on the 1998–2007 period, whereas the central government turn toward developmentalism occurred, as this chapter will show, from around 2006 onward.

3. Pearson 2005; Qin 2007; Mertha 2005a; Mertha and Zeng 2005.

4. Pierson 2000.

5. Kurlantzick 2016.

6. Heilmann and Shih 2013; Zheng 2004.

7. *2012 Report to Congress on China's WTO Compliance*, United States Trade Representative, December 2012, https://ustr.gov/sites/default/files/uploads/2012%20Report%20to%20Con gress%20-%20Dec%2021%20Final.pdf.

8. Heilmann and Shih 2013, 20.

9. Heilmann and Melton 2013, 585.

10. Zheng 2004.

11. Industry and information, human resources and social security, environmental protection, housing and urban-rural construction, and transportation. "China to Set Up Five New 'Super Ministries,' *Xinhua News*, March 11, 2008.

12. Yeo 2009, 738.

13. See, for example, one insider account of elite politics between Zhu Rongji and Jiang Zemin in 1998–1999 and the portrayal of constraints that Zhu seemingly had to operate under in the early years of his premiership (Zong 2002b).

14. The head of the General Office is a politically sensitive and important position to which leaders would only appoint officials whom they could trust to implement their policies.

15. Nathan and Gilley 2002, 101, 120.

16. Thelen 2004; Mahoney and Thelen 2010.

17. Pierson 2000.

18. Slater and Fenner 2011, 17.

19. Mann 1984.

20. Quoted in Zheng 2004, 93.

21. Jung 2006, 132. This concept is in turn based on Susan Shirk's concept of "reciprocal accountability" (Shirk 1993).

22. I am grateful to Tony Saich for inspiring this line of reasoning.

23. Shih, Adolph, and Liu 2012.

24. Zheng 2004; Li 2009; Shih 2008.

25. This is similar to Shih's (2008) distinction between generalists and technocrats.

26. Of course, leaders certainly have the ability to punish individual members of their network if a member is perceived to have betrayed the leader or no longer be loyal. The tension that this chapter focuses on is whether a leader is able to implement reforms that undermine the interests of the central bureaucracy when his network is also primarily located there—a very different proposition to sanctioning or purging individual members of a network.

27. The exact year in which one leadership period starts and ends is difficult to pinpoint, as party leadership positions are handed over at the Party Congress, which traditionally takes place in October or November. State leadership positions are handed over at the National People's Congress, which takes place a few months later in the spring of the following year. Hence, Hu and Wen handed their party leadership positions to Xi and Li in November 2012 and the state leadership positions in March 2013. For analytical simplicity, I mark each leadership period using the latter dates, when both party and state leadership positions have by and large been transferred.

28. Skocpol 1992.

29. The period of implementation largely ran from 2001 to 2005.

30. To clarify the scope conditions, the theory presented here does not apply to leaders such as Deng, who rose to power during the CCP's revolutionary days and in the volatile Mao era, when China's governance structure was far less institutionalized. The distinctions I make about careers in provinces versus the central government would therefore be less relevant to leaders from that era.

31. In Chinese, *Zhongguo Dang Zheng Ganbu Luntan, Zhonggong Zhongyang Dangxiao Xuebao,* and *Jingji Yanjiu Cankao.*

32. Interview B09, a government researcher.

33. Jung 2006, 125.

34. Zheng 2004, 102–6.
35. Zheng 2004, 97.
36. Zong 2002a.
37. The ministries of power, coal industry, metallurgical industry, machine building, chemical industry, and internal trade as well as the Textile Industry Council, the Light Industry Council, the General Company of Petroleum and Gas and the General Company of Chemical Industry (Zheng 2004, 98).
38. This section only provides a brief overview of the main developments, as detailed accounts can be found in Yang 2004, Zheng 2004, and Pearson 2005.
39. Yang 2004, 97.
40. Ibid., 39, 98.
41. Ibid., 98.
42. See Mertha 2005a on "soft centralization" and Yang 2004, 98.
43. Yang 2004, 103.
44. See Qin 2007 for an overview.
45. Interview G01, international organization official.
46. Jacob M. Schlesinger, "How China Swallowed the WTO," *Wall Street Journal*, November 1, 2017. China had been given an initial grace period to implement its WTO obligations, so it is possible that the increase in disputes filed in 2006 was also driven by the cessation of the grace period.
47. *2006 Report to Congress on China's WTO Compliance*, United States Trade Representative, December 11, 2006, https://ustr.gov/archive/assets/Document_Library/Reports_Publications /2006/asset_upload_file688_10223.pdf.
48. *2007 Report to Congress on China's WTO Compliance*, United States Trade Representative, December 11, 2007, https://ustr.gov/sites/default/files/asset_upload_file625_13692.pdf.
49. "China Issues Guidelines on Sci-Tech Development Program," *Xinhua News*, February 9, 2006.
50. Armaments, power generation and distribution, petrochemicals and oil, telecommunications, aviation, coal, and shipping.
51. E.g., equipment manufacturing, automotive, electronic information, construction, steel, nonferrous metals, chemicals, and high-tech industries. See Huanxin Zhao, "China Names Key Industries for Absolute State Control," *China Daily*, December 19, 2006; "Li Rongrong: China Speeds Up Strategic Adjustment of the State-Owned Economy" ("Li Rongrong: Zhongguo Jiakuai Guoyou Jingji Buju he Jiegou de Zhanlüe Tiaozheng"), State-owned Assets Supervision and Administration Commission of the State Council, September 9, 2005, https://web .archive.org/web/20130325024047/http://www.sasac.gov.cn/n1180/n1566/n259655/n260164 /11665146.html.
52. Suttmeier and Yao 2008.
53. Lardy 2014, 54.
54. "Wen Jiabao Chairs and Launches the Third Symposium for Developing Strategic Emerging Industries" ("Wen Jiabao Zhuchi Zhaokai Sanci Xinxing Zhanlüexing Chanye Fazhan Zuotan Hui"), Press Release, State Council General Office, September 22, 2009, http://www.gov.cn/ldhd/2009-09/22/content_1423493.htm.
55. Naughton 2009.
56. For details, see Yang 2004, 60–63; Lardy 2014, 49.
57. Guthrie, Xiao, and Wang 2015, 82.
58. Xiaoming Yu, "Central SOEs Down to 96 after Restructuring," *China Daily*, July 8, 2019.
59. Lardy 2014, 54.
60. Miller 2003, 6.
61. Yeo 2009; Heilmann and Shih 2013.
62. The Ministry of Information Industry (in charge of telecommunications and strategic electronics), the State Council Office on Informatization, parts of the NDRC in charge of technology and industrial management, and the National Defense Science and Technology Industry Office in charge of strategic military industries (Naughton 2008, 2009).
63. Interviews B30, a foreign businessperson, and B22, a central government official.
64. Interviews T04, a journalist, and B30.

65. Interviews B22, B30. The TPP was subsequently renamed the Comprehensive and Progressive Agreement for Trans-Pacific Partnership.

66. Interviews B09, a government researcher, and T04.

67. This is not to suggest that Jiang Zemin was in a politically strong position throughout his tenure. See, for example, MacFarquhar's assessment of Jiang's weakness in the early years of his tenure as general secretary and his consolidation of political power only after Deng's death in 1997 (MacFarquhar 2016). This latter period of strength from 1998 onward, however, is the relevant period that my analysis is focused on.

68. Information here is largely drawn from the official biographies of Chinese leaders on the People's Daily website (http://en.people.cn/). Additional sources are specified in the notes.

69. Song 2013, 429.

70. Shih 2008, 188.

71. Naughton 2003.

72. Lardy 2014, 45.

73. Interview with former US official, 2016.

74. Fewsmith 2001a; Pearson 2001.

75. Interview B18, a former Chinese government official.

76. Interview B04.

77. Pierson 1993.

78. Ibid.

79. For a detailed account of these disagreements, see Hopewell 2015.

80. Interview B18.

81. Permanent Mission of China to the WTO, *China in the WTO: Past, Present and Future*, World Trade Organization, 2011, https://www.wto.org/english/thewto_e/acc_e/s7lu_e.pdf.

82. Interview B22.

83. Interview B18.

84. Interview B22.

85. Pierson 1993; Conran and Thelen 2016.

86. Zheng 2004, 97.

87. Heilmann and Shih 2013.

88. Interview B18.

89. For a description of the WTO-plus commitments, see chapter 1 and Qin 2007.

90. Interview B18.

91. Interview B22.

92. It should be noted, however, that even under the Jiang-Zhu leadership, economic governance never turned completely toward full marketization or privatization. As Naughton (2002) has pointed out, Zhu Rongji's stance on economic governance has never been entirely market-driven or statist. Evidence from internal CCP documents as well as Zhu's public remarks suggest that he did not unequivocally support privatization and that he sought to use market mechanisms to strengthen rather than replace the core parts of the state sector. For example, in a 1994 internal discussion with members of the State Economic System Reform Commission, Zhu stressed that "it is possible to do a good job with SOEs; it is essential that we do so. This is an important issue that concerns the fate of the socialist system. We can do a good job with SOEs; we do not necessarily have to privatize" (State Economic System Reform Commission 1999, my translation). In remarks delivered in 1995 on the relationship between the party and the economy, Zhu emphasized that "party building needs to encircle the building of the economy, and the building of the economy needs to rely on party building" (Zhu 2011b, 149, my translation).

93. Naughton 2003.

94. Brown 2012, 17.

95. Fewsmith 2003.

96. Miller 2016.

97. Naughton 2003, 36.

98. Miller 2003.

99. Interview B26, an academic.

100. Naughton 2007b, 8.

101. Interview B18.

102. I discuss the implications of the trade war that emerged between the United States and China in 2018 in the concluding chapter.

103. "Proposed List of Ministries, Commissions of China's Cabinet after Reform," *Xinhua News*, March 13, 2018.

104. "China's Reform of Party, State Institutions Revolutionary: Senior Official," *Xinhua News*, March 13, 2018.

105. "CPC Central Committee Releases 'Plan on Deepening the Reform of Party and State Institutions,'" ("Zhonggong Zhongyang Yinfa 'Shenhua Dang he Guojia Jigou Gaige Fang'an'"), *Xinhua News*, March 21, 2018. The phrase in Chinese reads 党政军民学, 东西南北中, 党是领导一切的.

106. Naughton 2018, 1 (my emphasis).

107. Fang and Owen 2011; Chyzh 2014.

108. Skinner and Winckler 1969.

5. FDI and the Quest for National Champions

1. US Senate Committee on Small Business and Entrepreneurship, Project for Strong Labor Markets and National Development, *Made in China 2025 and the Future of American Industry*, 2019, www.rubio.senate.gov.

2. Walmsley, Hertel, and Ianchovichina 2006.

3. See Pandya 2016 for a recent comprehensive review.

4. See Harrison and Rodriguez-Clare 2010 for an overview.

5. Evans 1979; Pinto 2013; Mosley 2010.

6. Li and Resnick 2003; Jensen 2006; Simmons 2014; Büthe and Milner 2008.

7. See also Pandya 2014 on the factors affecting domestic demand for FDI.

8. Malesky 2008; Chen 2017.

9. This chapter, as with the rest of the book, engages primarily with China's manufacturing industries. As such, the "national champion" industries that I study here are in strategic manufacturing (automobiles, aviation, machinery, etc.) rather than important services sectors such as telecommunications and power generation.

10. Here, as with the rest of the book, I refer mainly to the incentives for party secretaries who hold the top political position at each level of administration rather than governors, mayors, or rank-and-file cadres for whom promotion may not be as important a factor in shaping their policy behavior.

11. See chapter 2 for a discussion of why growth rates dominate subnational promotional incentives despite an expansion of evaluation criteria over the years.

12. Zhu 2015, 111. This conference is attended by leaders from provinces and municipalities as well as central ministries and the People's Liberation Army.

13. Zhu 2015, 274–81.

14. Tariffs were cut from around 50 percent or more to about 25 percent for whole cars and 10 percent for automotive parts and from 3 percent to 0 percent for semiconductors.

15. In Chinese the strategy is called 市场换技术.

16. Pearson 1991.

17. There is less data for the semiconductor industry, as fewer regulations were issued in that period.

18. State Council of the People's Republic of China, "Circular on Strictly Controlling Production Points for Automobiles" ("Guanyu Yange Kongzhi Jiaoche Shengchandian de Tongzhi"), No. 82, 1988.

19. The "three large" firms referred to the central SOEs First Auto Works, Second Auto Works, and Shanghai Automotive Industry Corporation, while the "three small" referred to Beijing Jeep, Guangzhou-Peugeot, and Tianjin Auto Corporation (Huang 2002).

20. These restrictions proved difficult to maintain, however, and in the 1990s "under pressures from the military" two more firms were added as official production points (Chang An

and Guizhou Aviation Industry Corporation), making the policy "Three Large, Three Small and Two Minis" (Huang 2002, 546).

21. State Council of the People's Republic of China, "Circular on Issuing the Policy for Automotive Industry Development" ("Guowuyuan Guanyu Yinfa Qiche Gongye Chanye Zhengce de Tongzhi"), No. 17, 1994.

22. In Chinese the phrase is 国产化.

23. Anderson 2012. Starting with two centrally owned SOEs, First Auto Works established a joint venture with Volkswagen in 1991 and Dongfeng Motor Corporation with Citroen in 1992. General Motors established a joint venture with Shanghai Automotive Industry Corporation in 1997, followed by Honda with Guangzhou Auto Group in 1998.

24. National Development and Reform Commission, "Automotive Industry Development Policy" ("Qiche Chanye Fazhan Zhengce"), No. 8, 2004.

25. Against central policy, several successful private auto companies had emerged in the past decade such as Great Wall, Geely, and BYD. In 2007, the NDRC approved three private carmakers (alongside two local SOEs) to produce Chinese-branded passenger cars, possibly indicative of an emerging willingness to support private firms (Anderson 2012).

26. NDRC, "Automotive Industry Development Policy," Chapter 10, Article 44.

27. "China to Relax Foreign Car Venture Limit in Orderly Manner," *Bloomberg News*, April 24, 2017.

28. Ministry of Industry and Information Technology, National Development and Reform Commission, Ministry of Science and Technology, "Circular Regarding the Medium to Long-term Plan for Automotive Industry Development" ("San Buwei Guanyu Yinfa Qiche Chanye Zhongchangqi Fazhan Guihua de Tongzhi"), April 6, 2017.

29. Keith Bradsher, "China Loosens Foreign Auto Rules, in Potential Peace Offering to Trump," *New York Times*, April 17, 2018.

30. Chen 2005.

31. Shanghai Philips Semiconductor in 1988 (with Philips NV), Shougang NEC Electronics in 1991 (with Japan's NEC Corporation), and Shanghai Belling Microelectronics Manufacturing Corporation in 1994 (with Alcatel Bell). The other two backbone enterprises were Huajing Electronic Enterprise Group based in Jiangxi (one of China's largest semiconductor SOEs) and Huawei Electronics in Zhejiang (Pecht 2006).

32. Jefferson and Ouyang 2014; Pecht 2006.

33. Interview B26.

34. The figure captures changes in the major types of enterprise registration but is not exhaustive (there are nineteen different registration classifications for firms in China). In addition, the joint venture and WFOE firms include firms from Hong Kong, Taiwan, and Macau as well as foreign firms.

35. Interviews B26 and B30. This privileged status was compared to that of the eldest son in a family.

36. PricewaterhouseCoopers, "China's Impact on the Semiconductor Industry: 2016 Update," 2017.

37. Interview S05.

38. Interview SZ02.

39. Chou, Chang, and Li 2014.

40. Interview S06.

41. PricewaterhouseCoopers, "China's Impact on the Semiconductor Industry, 2004 Update," 2004; PricewaterhouseCoopers, "China's Impact on the Semiconductor Industry, 2008 Update," 2008.

42. Interview B26.

43. Chou, Chang, and Li 2014, 5657, 5665.

44. Interview S06.

45. Chou et al. 2011.

46. Chen 2017.

47. Article II on the Schedule of Concessions, Article III on National Treatment, Article XI on the Elimination of Quantitative Restrictions, and Article XIII on the Non-discriminatory

Administration of Quantitative Restrictions as well as the agreements on subsidies and countervailing measures and trade-related investment measures and the Protocol of Accession.

48. "WTO Dispute Settlement: One Page Summaries; China—Auto Parts," World Trade Organization, 2017, https://www.wto.org/english/tratop_e/dispu_e/cases_e/1pagesum_e/ds339 sum_e.pdf.

49. Interview B30.

50. Ibid.

51. Interview B28.

52. *China's Approval Process for Inbound Foreign Direct Investment: Impact on Market Access, National Treatment and Transparency*, US Chamber of Commerce, 2012, https://www.uschamber.com/sites/default/files/documents/files/020021_China_InboundInvestment_Cvr.pdf.

53. Interview B26.

54. Interview SZ01.

55. See OECD 2008 for an overview of this plan and the priority placed on indigenous innovation.

56. State Council of the People's Republic of China, "Circular on Issuing the Twelfth Five-Year Development Plan for National Strategic Emerging Industries ("Guowuyuan Guanyu Yinfa 'Shi'er Wu' Guojia Zhanlüexing Chanye Fazhan Guihua de Tongzhi"), No. 28, July 20, 2012, http://www.gov.cn/zwgk/2012-07/20/content_2187770.htm.

57. Interview S05.

58. Interviews S05, B26, and T05.

59. Chuin-Wei Yap and Paul Mozur, "China Aims to Create Electronics Giants," *Wall Street Journal*, January 22, 2013.

60. Interview T05.

61. State Council of the People's Republic of China, "Outline for Spurring the Development of the National Integrated Circuit Industry" ("Guojia Jicheng Dianlu Chanye Fazhan Tuijin Gangyao"), June, 2014, www.pkulaw.cn.

62. Bien Perez, "China's Chip Industry Bolstered by Acquisitions Worth US$6.61 Billion by Government-Backed Fund," *South China Morning Post*, March 30, 2016.

63. Miyoung Kim and Makiko Yamazaki, "China's Chip Crusade to Intensify Sector's Record M&A Run," *Reuters*, November 25, 2015; "Chips on their Shoulders," *The Economist*, January 23, 2016, https://www.economist.com/business/2016/01/23/chips-on-their-shoulders.

64. "Tsinghua Leads Creation of $2.8 Billion Chinese Chip Giant," Bloomberg News, July 26, 2016, https://www.bloomberg.com/news/articles/2016-07-26/tsinghua-unigroup-xmc-merge-chip-assets-to-create-chinese-giant.

65. Interview B26.

66. State Council of the People's Republic of China, "Circular of the State Council on Issuing 'Made in China 2025,'" ("Guowuyuan Guanyu Yinfa 'Zhongguo Zhizao 2025' de Tongzhi") No. 28, May 8, 2015, http://www.gov.cn/zhengce/content/2015-05/19/content_9784.htm.

67. Interview S11.

68. Interview S04.

69. Ian King, "China Has Big Plans for Homegrown Chips," Bloomberg News, June 25, 2015, https://www.bloomberg.com/news/articles/2015-06-25/china-has-big-plans-for-homegrown-chips.

70. King, "China Has Big Plans," June 25, 2015.

71. Interview T05.

72. "Secretary Ross Announces Activation of ZTE Denial Order in Response to Repeated False Statements to the U.S. Government," United States Department of Commerce, April 16, 2018, https://www.commerce.gov/news/press-releases/2018/04/secretary-ross-announces-activation-zte-denial-order-response-repeated.

73. "Xi Jinping: Indigenous Innovation Advances the Construction of Cyber Power" ("Xi Jinping: Zizhu Chuangxin Tuijin Wangluo Qiangguo Jianshe"), *Xinhua News*, April 21, 2018.

74. "Executive Order on Securing the Information and Communications Technology and Services Supply Chain," The White House, May 15, 2019, https://www.whitehouse.gov/presidential-actions/executive-order-securing-information-communications-technology-services

-supply-chain/; "Bureau of Industry and Security Entity List," United States Department of Commerce, https://www.bis.doc.gov/index.php/policy-guidance/lists-of-parties-of-concern/entity-list.

75. Jun Mai and Amanda Lee, "Xi Urges Self-Reliance as Long-Term Challenges from US Grow," *South China Morning Post*, May 29, 2019.

76. "President Donald J. Trump Is Committed to Safeguarding America's Vital Communications Networks and Securing 5G Technology," The White House, March 12, 2020, https://www.whitehouse.gov/briefings-statements/president-donald-j-trump-committed-safeguarding-americas-vital-communications-networks-securing-5g-technology/; Ana Swanson, "U.S. Delivers Another Blow to Huawei with New Tech Restrictions," *New York Times*, May 15, 2020; "FCC Designates Huawei and ZTE as National Security Threats," June 30, 2020, Federal Communications Commission, https://www.fcc.gov/document/fcc-designates-huawei-and-zte-national-security-threats.

77. This is a high figure compared to the rest of the economy. In 2007, only five industries had a higher share of central SOE output: tobacco and nuclear fuel processing, which are almost entirely state-owned, and electronic/electrical machinery equipment manufacturing, aircraft manufacturing, and petroleum refining—all highly strategic industries.

78. I follow the approach adopted by Zhu 2012 in measuring the degree to which the central government encourages investment in an industry.

79. One challenge to consider in the estimation is that relative policy activism can both be a result of and have an effect on industry characteristics. For example, if more active central policies lead to faster growing industries, then the dependent variable might affect some of the covariates such as output and employment shares. To deal with this issue, the estimations are run with all the explanatory and control variables lagged by a year. All estimations additionally include industry fixed effects, to capture time-invariant industry characteristics that might affect policy activism, as well as year fixed effects to capture year-specific shocks or trends affecting all industries over time that might have an effect on policy activism. The standard deviation of tariff rates for an industry are also included as a control to account for the aggregation of product-level tariff rates up to the broader industry level. Standard errors are clustered by industry.

Conclusion

1. *National Security Strategy of the United States of America*, The White House, December 2017, https://www.whitehouse.gov/wp-content/uploads/2017/12/NSS-Final-12-18-2017-0905.pdf..

2. Evans 1995.

3. Hall and Soskice 2001.

4. Thelen 2012.

5. Garrett 1998; Clift and Woll 2012.

6. Goldstein 1988; Shaffer 2015.

7. Mercurio and Tyagi 2012; Gao 2011; Kennedy and Cheng 2012; Li 2012.

8. Tomz 1997; Schoppa 1993.

9. EU membership involves satisfying a thick set of commitments, from reforming one's legal administration to conform with the body of European law (the Acquis Communautaire) to meeting the Copenhagen criteria of democracy, the rule of law, protection for human rights, and more as well as fiscal rules contained in the Growth and Stability Pact.

10. Way and Levitsky 2007.

11. Levitz and Pop-Eleches 2010.

12. Li 2012.

13. Vu-Thanh 2017.

14. Just as Mertha (2005b) illuminates how the efficacy of external pressure on intellectual property rights protection is shaped by China's complex domestic bureaucratic politics and local-level coalitions, this study underscores how globalization has transformed the distribution of power within China's fragmented bureaucracy.

15. Dittmer and Wu 1995; Shih 2008; Shih, Adolph, and Liu 2012.

16. Tsai 2007.

17. See for example Moore 2002.
18. These important studies include Yang 2004; Mertha 2005a; and Pearson 2007.
19. Hsueh 2011.
20. Steinfeld 2012; Johnston 2008.
21. Mahoney and Thelen 2010; Krasner 1984.
22. Naughton 2007a; Rawski 1999; Tsai 2002.
23. Heilmann 2007; Ang 2016.
24. Skinner and Winckler 1969. Huang (2008) argues that 1980s China was a period of market liberalism favoring rural growth and private entrepreneurship, while the 1990s involved a great reversal back to statism and policies favoring urban growth. Shih (2008), in explaining inflation control, argues that the party delegates power to the provinces during periods of prosperity, then transfers that authority to central technocrats during periods of inflation.
25. Heilmann 2007.
26. Even in the security arena, however, Taylor Fravel finds that the CCP is able to update its military strategy to address systemic changes in the war-fighting environment only when the party is united (Fravel 2019). That is, even in the matter of national security, issues of internal coordination affect the CCP's ability to respond to shifts in the international system.
27. Cimino-Isaacs 2016.
28. Wade 2014; Schrank and Whitford 2009; Block 2008.
29. See information on the Michigan-China Innovation Center at www.michigan.org.
30. Indeed, US state government initiatives to boost economic activity "resemble the national industrial policies of France and Japan both in terms of the structure of the underlying economic plans and in their programmatic emphasis" (Eisinger 1990, 509).
31. Smith 2005.
32. "European Commission Industrial Policy Strategy," European Commission, 2017, http://europa.eu/rapid/attachment/IP-17-3185/en/Making%20Europe_s%20industry%20stronger_Key%20Commission.
33. Javier Espinoza, Arthur Beesley, Tim Bradshaw, and Aime Williams, "Apple Wins Landmark Court Battle with EU over €14.3bn of Tax Payments," *Financial Times*, July 15, 2020.
34. Rogowski 1987b ; Lohmann and O'Halloran 1994; Schiller 1999.
35. Thielmann 1999.
36. For example, Brazilian subnational governments are engaged in fierce competition with each other to attract foreign investment, leading them to engage in market-shaping developmentalism by extending tax benefits and other forms of public support to foreign firms (Rodrigues 2003).
37. Malesky 2008.
38. Autor, Dorn, and Hanson 2016.
39. NAFTA was renegotiated and renamed the US-Mexico-Canada Agreement in 2018.
40. Tiefer 1998, 47.
41. The EC has exclusive competency over trade in goods, trade in electronically transmitted services and trade rules, and mixed competency (i.e., shared with member states) over trade-related issues outside of these specific areas.
42. Steinberg 2006, 355.
43. Hudec 1993; Petersmann 1997.
44. Hale 2004.
45. Leblond and Fafard 2014.
46. "AbitibiBowater Inc. v. Government of Canada," Global Affairs Canada, 2013, https://www.international.gc.ca/trade-agreements-accords-commerciaux/topics-domaines/disp-diff/AbitibiBowater.aspx?lang=eng.
47. Peter Berg, "The Canada-U.S. Softwood Lumber Dispute," Canada Library of Parliament, Publication No. TIPS-98E, Parliamentary Information and Research Service, Library of Parliament, Ottawa, June 10, 2004.
48. Tamayo 2001.
49. Geddes 1999; Levitsky and Way 2002.

References

Adhikari, Ramesh, and Yongzheng Yang. 2002. "What Will WTO Membership Mean for China and Its Trading Partners?" *Finance and Development* 39, no. 3, https://www.imf.org/external/pubs/ft/fandd/2002/09/adhikari.htm.

Amsden, Alice H., and Takashi Hikino. 2000. "The Bark Is Worse Than the Bite: New WTO Law and Late Industrialization." *Annals of the American Academy of Political and Social Science* 570, no. 1: 104–14.

Anderson, G. E. 2012. *Designated Drivers*. Singapore: Wiley.

Ang, Yuen Yuen. 2012. "Counting Cadres: A Comparative View of the Size of China's Public Employment." *China Quarterly* 211: 676–96.

——. 2016. *How China Escaped the Poverty Trap*. Cornell Studies in Political Economy. Ithaca, NY: Cornell University Press.

Autor, David H., David Dorn, and Gordon H. Hanson. 2016. "The China Shock: Learning from Labor-Market Adjustment to Large Changes in Trade." *Annual Review of Economics* 8: 205–40.

Bachman, David. 1992. "The Limits on Leadership in China." *Asian Survey* 32, no. 11: 1046–62.

Barton, John H. 2006. *The Evolution of the Trade Regime: Politics, Law, and Economics of the GATT and the WTO*. Princeton, NJ: Princeton University Press.

Baum, Richard, and Alexei Shevchenko. 1999. "The 'State of the State.'" In *The Paradox of China's Post-Mao Reforms*, ed. Merle Goldman and Roderick MacFarquhar, 333–60. Cambridge, MA: Harvard University Press.

Blei, David M., and John D. Lafferty. 2007. "A Correlated Topic Model of Science." *Annals of Applied Statistics* 1, no. 1: 17–35.

Block, Fred. 2008. "Swimming against the Current: The Rise of a Hidden Developmental State in the United States." *Politics & Society* 36, no. 2: 169–206.

Brown, Kerry. 2012. *Hu Jintao: China's Silent Ruler*. Hackensack, NJ: World Scientific.

Büthe, Tim, and Helen V. Milner. 2008. "The Politics of Foreign Direct Investment into Developing Countries: Increasing FDI through International Trade Agreements?" *American Journal of Political Science* 52, no. 4: 741–62.

Campbell, Kurt M., and Ely Ratner. 2018. "The China Reckoning: How Beijing Defied American Expectations." *Foreign Affairs* 97, no. 2 (March–April), https://www.foreignaffairs.com/articles/united-states/2018-02-13/china-reckoning.

Carpenter, Daniel P. 2001. *The Forging of Bureaucratic Autonomy: Reputations, Networks, and Policy Innovation in Executive Agencies, 1862–1928.* Princeton Studies in American Politics. Princeton, NJ: Princeton University Press.

Chang, Ha-Joon. 2002. *Kicking Away the Ladder: Development Strategy in Historical Perspective.* London: Anthem.

Chen, Bingcai. 1996. "Characteristics and Debates over Korea's Large Conglomerates" (Hanguo Daqiye de Tezheng ji Zhengyi), *Review of Economic Research (Jingji Yanjiu Cankao)*, no. 58: 34–38.

Chen, Ling. 2005. "Institution, Elite and Consensus: The Policy Process of China's Semiconductor Industrial Policies." PhD diss., Tsinghua University.

——. 2017. "Grounded Globalization: Foreign Capital and Local Bureaucrats in China's Economic Transformation." *World Development* 98: 381–99.

Chou, Tsu Lung, et al. 2011. "Global Linkages, the Chinese High-Tech Community and Industrial Cluster Development: The Semiconductor Industry in Wuxi, Jiangsu." *Urban Studies* 48, no. 14: 3019–42.

Chou, Tsu-Lung, Jung-Ying Chang, and Te-Chuan Li. 2014. "Government Support, FDI Clustering and Semiconductor Sustainability in China: Case Studies of Shanghai, Suzhou and Wuxi in the Yangtze Delta." *Sustainability* 6, no. 9: 5655–81.

Chung, Jae Ho. 2016. *Centrifugal Empire: Central-Local Relations in China.* New York: Columbia University Press.

Chyzh, Olga. 2014. "Can You Trust a Dictator: A Strategic Model of Authoritarian Regimes' Signing and Compliance with International Treaties." *Conflict Management and Peace Science* 31, no. 1: 3–27.

Cimino-Isaacs, Cathleen. 2016. *Assessing the Trans-Pacific Partnership,* Vol. 2, *Innovations in Trading Rules.* Washington, DC: Peterson Institute for International Economics.

Clift, Ben, and Cornelia Woll. 2012. "Economic Patriotism: Reinventing Control over Open Markets." *Journal of European Public Policy* 19, no. 3: 307–23.

Conran, James, and Kathleen Thelen. 2016. "Institutional Change." In *The Oxford Handbook of Historical Institutionalism,* ed. Orfeo Fioretos, Tulia G. Falleti, and Adam Sheingate, 51–70. Oxford: Oxford University Press.

Cornick, Jorge. 2014. "Transforming Economies: Making Industrial Policy Work for Growth, Jobs and Development." *International Labour Review* 153, no. 3: 509–11.

Davis, Christina. 2012. *Why Adjudicate? Enforcing Trade Rules in the WTO.* Princeton, NJ: Princeton University Press.

Davis, Christina, and Meredith Wilf. 2011. "Joining the Club: Accession to the GATT/WTO." *Journal of Politics* 79, no. 3: 964–78.

Devereaux, Charan, Robert Z. Lawrence, and Michael Watkins. 2006. "China's Accession: The 1999 US-China Bilateral Agreement and the Battle for PNTR." In *Case Studies in US Trade Negotiation,* Vol. 1, *Making the Rules,* 241–300. Washington, DC: Peterson Institute for International Economics.

DeWoskin, Kenneth J. 2001. "The WTO and the Telecommunications Sector in China." *China Quarterly* 167: 630–54.

Dittmer, Lowell, and Yu-Shan Wu. 1995. "The Modernization of Factionalism in Chinese Politics." *World Politics* 47, no. 4: 467–94.

Donaldson, John. 2016. *Assessing the Balance of Power in Central-Local Relations in China.* Routledge Contemporary China Series. London: Routledge.

Edin, Maria. 2003. "State Capacity and Local Agent Control in China: CCP Cadre Management from a Township Perspective." *China Quarterly* 173: 35–52.

Eisinger, Peter. 1990. "Do the American States Do Industrial Policy?" *British Journal of Political Science* 20, no. 4: 509–35.

Evans, Peter B. 1979. *Dependent Development: The Alliance of Multinational, State, and Local Capital in Brazil.* Princeton, NJ: Princeton University Press.

——. 1995. *Embedded Autonomy: States and Industrial Transformation.* Princeton, NJ: Princeton University Press.

Fang, Songying, and Erica Owen. 2011. "International Institutions and Credible Commitment of Non-Democracies." *Review of International Organizations* 6, no. 2: 141–62.

Feenstra, Robert C., and Akira Sasahara. 2018. "The 'China Shock,' Exports and US Employment: A Global Input-Output Analysis." *Review of International Economics* 26, no. 5: 1053–83.

Feng, Hui. 2006. *The Politics of China's Accession to the World Trade Organization: The Dragon Goes Global.* New York: Routledge.

Fewsmith, Joseph. 2000. "The Politics of China's Accession to the WTO." *Current History* 99, no. 638: 268–73.

——. 2001a. "The Political and Social Implications of China's Accession to the WTO." *China Quarterly* 167: 573–91.

——. 2001b. "Tiananmen and the Conservative Critique of Reform." In *China since Tiananmen: The Politics of Transition,* 21–43. Cambridge: Cambridge University Press.

——. 2003. "The Sixteenth National Party Congress: The Succession That Didn't Happen." *China Quarterly* 173: 1–16.

Fravel, M. Taylor. 2019. *Active Defense: China's Military Strategy since 1949.* Princeton, NJ: Princeton University Press.

Frieden, Jeffry. 1991. "Invested Interests: The Politics of National Economic Policies in a World of Global Finance." *International Organization* 45, no. 4: 425–51.

Gao, Henry S. 2007. "China's Participation in the WTO: A Lawyer's Perspective." *Singapore Year Book of International Law* 11: 41–74.

Garrett, Geoffrey. 1998. "Global Markets and National Politics: Collision Course or Virtuous Circle?" *International Organization* 52, no. 4: 787–824.

——. 2001. "Globalization and Government Spending around the World." *Studies in Comparative International Development* 35, no. 4: 3–29.

Geddes, Barbara. 1999. "What Do We Know about Democratization after Twenty Years?" *Annual Review of Political Science* 2, no. 1: 115–44.

Gewirtz, Julian. 2017. *Unlikely Partners: Chinese Reformers, Western Economists, and the Making of Global China.* Cambridge, MA: Harvard University Press.

Goldstein, Judith. 1988. "Ideas, Institutions, and American Trade Policy." *International Organization* 42, no. 1: 179–217.

——. 2012. "Trade Liberalization and Domestic Politics." In *Oxford Handbook on the World Trade Organization,* ed. Martin Daunton, Amrita Narlikar, and Robert Stern, 64–82. Oxford: Oxford University Press.

Goldstein, Judith, and Lisa Martin. 2000. "Legalization, Trade Liberalization, and Domestic Politics: A Cautionary Note." *International Organization* 54, no. 3: 603–32.

Goldstein, Judith, Douglas Rivers, and Michael Tomz. 2007. "Institutions in International Relations: Understanding the Effects of the GATT and the WTO on World Trade." *International Organization* 61, no. 1: 37–67.

Gourevitch, Peter. 1978. "The Second Image Reversed: The International Sources of Domestic Politics." *International Organization* 32, no. 4: 881–912.

Grossman, Gene, and Elhanan Helpman. 1994. "Protection for Sale." *American Economic Review* 84, no. 4: 833–50.

Guthrie, Doug, Zhixing Xiao, and Junmin Wang. 2015. "Stability, Asset Management, and Gradual Change in China's Reform Economy." In *State Capitalism, Institutional Adaptation, and the Chinese Miracle,* ed. Barry Naughton and Kellee Tsai, 75–101. New York: Cambridge University Press.

Hale, Geoffrey E. 2004. "Canadian Federalism and the Challenge of North American Integration." *Canadian Public Administration* 47, no. 4: 497–524.

Hall, Peter A., and David Soskice. 2001. *Varieties of Capitalism: The Institutional Foundations of Comparative Advantage.* Oxford: Oxford University Press.

Harpaz, Marcia Don. 2011. "China's WTO Compliance-Plus Anti-Dumping Policy." *Journal of World Trade* 45, no. 4: 727–66.

Harrison, Ann, and Andres Rodriguez-Clare. 2010. "Chapter 63 Trade, Foreign Investment, and Industrial Policy for Developing Countries." In *Handbook of Development Economics*, Vol. 5, ed. Dani Rodrik and Mark Rosenzweig, 4039–214. Oxford, UK: North Holland.

Harwit, Eric. 2001. "The Impact of WTO Membership on the Automobile Industry in China." *China Quarterly* 167: 655–70.

Hausmann, Ricardo, Dani Rodrik, and Charles Sabel. 2008. "Reconfiguring Industrial Policy: A Framework with an Application to South Africa." CID Working Papers at Harvard University, https://ideas.repec.org/p/cid/wpfacu/168.html.

Heilmann, Sebastian. 2007. "Policy Experimentation in China's Economic Rise." *Studies in Comparative International Development* 43, no. 1: 1–26.

Heilmann, Sebastian, and Oliver Melton. 2013. "The Reinvention of Development Planning in China, 1993–2012." *Modern China* 39, no. 6: 580–628.

Heilmann, Sebastian, and Lea Shih. 2013. "The Rise of Industrial Policy in China, 1978–2012." Harvard-Yenching Institute Working Paper Series, https://www.harvard-yenching.org/research/hyi-working-paper-series-sebastian-heilmann-and-lea-shih/.

Heilmann, Sebastian, Lea Shih, and Andreas Hofem. 2013. "National Planning and Local Technology Zones: Experimental Governance in China's Torch Programme." *China Quarterly* 216: 896–919.

Hiscox, Michael J. 2001. "Class versus Industry Cleavages: Inter-Industry Factor Mobility and the Politics of Trade." *International Organization* 55, no. 1: 1–46.

Hoekman, Bernard M., and Petros C. Mavroidis. 2000. "WTO Dispute Settlement, Transparency and Surveillance." *World Economy* 23, no. 4: 527–42.

Hopewell, Kristen. 2015. "Different Paths to Power: The Rise of Brazil, India and China at the World Trade Organization." *Review of International Political Economy* 22, no. 2: 311–38.

Hsueh, Roselyn. 2011. *China's Regulatory State: A New Strategy for Globalization.* Cornell Studies in Political Economy. Ithaca, NY: Cornell University Press.

Huan, Guoyu. 1996. "The Experience and Lessons of Macroeconomic Management in the High-Speed Growth Periods in Korea and Japan" (Han, Ri Liangguo Jingji Gaosu Zengzhang Shiqi Hongguan Jingji Guanli de Jingyan Ji Qishi). *Review of Economic Research* (*Jingji Yanjiu Cankao*) no. 94: 30–40.

Huang, Yasheng. 2002. "Between Two Coordination Failures: Automotive Industrial Policy in China with a Comparison to Korea." *Review of International Political Economy* 9, no. 3: 538–73.

——. 2008. *Capitalism with Chinese Characteristics: Entrepreneurship and the State.* New York: Cambridge University Press.

Hudec, Robert E. 1993. *Enforcing International Trade Law: The Evolution of the Modern GATT Legal System.* Salem, NH: Butterworth Legal Publishers.

Hurst, William. 2009. *The Chinese Worker after Socialism.* New York: Cambridge University Press.

Ikenberry, G. John. 2009. "Liberal Internationalism 3.0: America and the Dilemmas of Liberal World Order." *Perspectives on Politics* 7, no. 1: 71–87.

Jaros, Kyle A., and Yeling Tan. 2020. "Provincial Power in a Centralizing China: The Politics of Domestic and International 'Development Space.'" *China Journal* 83, no. 1: 79–104.

Jefferson, Gary H., and Miao Ouyang. 2014. "FDI Spillovers in China: Why Do the Research Findings Differ So Much?" *Journal of Chinese Economic and Business Studies* 12, no. 1: 1–27.

Jensen, Nathan M. 2006. *Nation-States and the Multinational Corporation: A Political Economy of Foreign Direct Investment*. Princeton, NJ: Princeton University Press.

Johnson, Chalmers. 1982. *MITI and the Japanese Miracle: The Growth of Industrial Policy, 1925–1975*. Stanford, CA: Stanford University Press.

Johnston, Alastair I. 2008. *Social States: China in International Institutions, 1980–2000*. Princeton Studies in International History and Politics. Princeton, NJ: Princeton University Press.

Jung, Joo-Youn. 2006. "When Nonliberal Economies Meet Globalization: The Transformation of Interventionist States in East Asia." PhD diss., Stanford University.

Kennedy, Scott. 2009. *The Business of Lobbying in China*. Cambridge, MA: Harvard University Press.

Kennedy, Scott, and Shuaihua Cheng. 2012. *From Rule Takers to Rule Makers: The Growing Role of Chinese in Global Governance*. Bloomington, IN: Research Center for Chinese Politics & Business, Indiana University.

Kennedy, Scott, and Christopher Johnson. 2016. "*Perfecting China, Inc. China's 13th Five-Year Plan*." Report of the CSIS Freeman Chair in China Studies. Washington, DC: Center for Strategic and International Studies.

Krasner, Stephen D. 1984. "Approaches to the State: Alternative Conceptions and Historical Dynamics." *Comparative Politics* 16, no. 2: 223–46.

Krasner, Stephen D., and Jeremy M. Weinstein. 2014. "Improving Governance from the Outside In." *Annual Review of Political Science* 17: 123–45.

Kurlantzick, Joshua. 2016. *State Capitalism: How the Return of Statism Is Transforming the World*. New York: Oxford University Press.

Landry, Pierre F. 2008. *Decentralized Authoritarianism in China: The Communist Party's Control of Local Elites in the Post-Mao Era*. New York: Cambridge University Press.

Landry, Pierre F., Xiaobo Lü, and Haiyan Duan. 2018. "Does Performance Matter? Evaluating Political Selection along the Chinese Administrative Ladder." *Comparative Political Studies* 51, no. 8: 1074–105.

Lardy, Nicholas R. 2014. *Markets over Mao: The Rise of Private Business in China*. Washington, DC: Peterson Institute for International Economics.

Lau, Lawrence J., Yingyi Qian, and Gerard Roland. 2000. "Reform without Losers: An Interpretation of China's Dual-Track Approach to Transition." *Journal of Political Economy* 108, no. 1: 120–43.

Leblond, Patrick, and Patrick Fafard. 2014. "Canadian Federalism and International Trade: A Small Step While Waiting for the Giant Leap." In *The State of the Federation, 2011: The Changing Federal Environment; Rebalancing Roles*, ed. Nadia Verrelli, 153–68. Kingston, Ontario: Institute of Intergovernmental Relations.

Levitsky, Steven, and Lucan Way. 2002. "The Rise of Competitive Authoritarianism." *Journal of Democracy* 13, no. 2: 51–65.

Levitz, Philip, and Grigore Pop-Eleches. 2010. "Why No Backsliding? The European Union's Impact on Democracy and Governance before and after Accession." *Comparative Political Studies* 43, no. 4: 457–85.

Levy, Jonah D. 2006. *The State after Statism: New State Activities in the Age of Liberalization*. Cambridge, MA: Harvard University Press.

Li, Cheng. 2009. "China's Team of Rivals." *Foreign Policy* 171 (March–April): 88–93.

Li, Hongbin, and Li-An Zhou. 2005. "Political Turnover and Economic Performance: The Incentive Role of Personnel Control in China." *Journal of Public Economics* 89, no. 9: 1743–62.

Li, Qinghua. 2001. "How to Exert the Role of Government in the Context of Economic Globalization: Interview with State Council Development Reform Center Deputy Director Chen Qingtai" (Jingji Quanqiuhua Xia Beijing Ruhe Fahui Zhengfu Zuoyong: Fang Guowuyuan Fazhan Yanjiuyuan Fuzhuren Chen Qingtai). *Chinese Cadres Tribune*, no. 5 (2001): 3–5.

Li, Quan, and Adam Resnick. 2003. "Reversal of Fortunes: Democratic Institutions and Foreign Direct Investment Inflows to Developing Countries." *International Organization* 57, no. 1: 175–211.

Li, Xiaojun. 2012. "Understanding China's Behavioral Change in the WTO Dispute Settlement System." *Asian Survey* 52, no. 6: 1111–37.

Liang, Liping. 2001. "Strengthening the Building of the Ruling Party with the Spirit of Reform: Interview with Central Party School Deputy Principal Li Junru" (Yi Gaige De Jingshen Jiaqiang Zhizheng Dang Jianshe: Fang Zhonggong Zhongyang Dangxiao Fuxiaozhang Li Junru). *Chinese Cadres Tribune*, no. 9: 4–8.

Lieberthal, Kenneth, and Michel Oksenberg. 1988. *Policy Making in China: Leaders, Structures, and Processes.* Princeton, NJ: Princeton University Press.

Lohmann, Susanne, and Sharyn O'Halloran. 1994. "Divided Government and U.S. Trade Policy: Theory and Evidence." *International Organization* 48, no. 4: 595–632.

Ma, Licheng. 2015. *Leading Schools of Thought in Contemporary China.* Singapore: World Scientific.

MacFarquhar, Roderick. 2016. "Leadership Styles at the Party Centre: From Mao Zedong to Xi Jinping." In *China's Core Executive Leadership Styles, Structures and Processes under Xi Jinping*, ed. Sebastian Heilmann and Matthias Stepan, 14–17. Berlin: Mercator Institute for China Studies.

Mahoney, James, and Kathleen Thelen. 2010. *Explaining Institutional Change: Ambiguity, Agency, and Power.* Cambridge: Cambridge University Press.

Malesky, Edmund J. 2008. "Straight Ahead on Red: How Foreign Direct Investment Empowers Subnational Leaders." *Journal of Politics* 70, no. 1: 97–119.

Mann, Michael. 1984. "The Autonomous Power of the State: Its Origins, Mechanisms and Results." *European Journal of Sociology* 25, no. 2: 185–213.

Mansfield, Edward D., Helen V. Milner, and B. Peter Rosendorff. 2000. "Free to Trade: Democracies, Autocracies, and International Trade." *American Political Science Review* 94, no. 2: 305–21.

Mei, Ciqi, and Margaret M. Pearson. 2014. "Killing a Chicken to Scare the Monkeys? Deterrence Failure and Local Defiance in China." *China Journal* 72 (July): 75–97.

Mercurio, Bryan, and Mitali Tyagi. 2012. "China's Evolving Role in WTO Dispute Settlement: Acceptance, Consolidation and Activation." In *European Yearbook of International Economic Law*, Vol. 3, ed. Christoph Herrmann and Jörg Philipp Terhechtev, 89–124. Berlin, Heidelberg: Springer.

Mertha, Andrew. 2005a. "China's 'Soft' Centralization: Shifting Tiao/Kuai Authority Relations." *China Quarterly* 184: 791–810.

——. 2005b. *The Politics of Piracy: Intellectual Property in Contemporary China.* Ithaca, NY: Cornell University Press.

——. 2009. "'Fragmented Authoritarianism 2.0': Political Pluralization in the Chinese Policy Process." *China Quarterly* 200: 995–1012.

Mertha, Andrew, and Ka Zeng. 2005. "Political Institutions, Resistance and China's Harmonization with International Law." *China Quarterly* 182: 319–37.

Miller, Alice. 2003. "The 10th National People's Congress and China's Leadership Transition." *China Leadership Monitor* 7 (Summer). https://www.hoover.org/sites/default/files/uploads/documents/clm7_lm.pdf.

——. 2016. "'Core' Leaders, 'Authoritative Persons,' and Reform Pushback." *China Leadership Monitor* 50: 2–3.

Moore, Thomas Geoffrey. 2002. *China in the World Market: Chinese Industry and International Sources of Reform in the Post-Mao Era.* Cambridge: Cambridge University Press.

Mosley, Layna. 2010. *Labor Rights and Multinational Production.* New York: Cambridge University Press.

Nathan, Andrew J., and Bruce Gilley, eds. 2002. *China's New Rulers: The Secret Files.* New York: New York Review Books.

Nathan, Andrew J. 2003. "Authoritarian Resilience." *Journal of Democracy* 14, no. 1: 6–17.

Naughton, Barry. 2002. "Zhu Rongji: The Twilight of a Brilliant Career." *China Leadership Monitor* 1 (Winter): 1–10.

———. 2003. "The Emergence of Wen Jiabao." *China Leadership Monitor* 6 (Winter): 36–47.

———. 2007a. *The Chinese Economy: Transitions and Growth*. Cambridge, MA: MIT Press.

———. 2007b. "Strengthening the Center, and Premier Wen Jiabao." *China Leadership Monitor* 21 (Summer): 1–10.

———. 2008. "A New Team Faces Unprecedented Economic Challenges." *China Leadership Monitor* 26 (Fall): 36–87.

———. 2009. "China's Emergence from Economic Crisis." *China Leadership Monitor* 29 (Summer): 1–10.

———. 2018. "Xi's System, Xi's Men: After the March 2018 National People's Congress." *China Leadership Monitor* 56 (Spring) 2018. https://www.hoover.org/sites/default/files /research/docs/clm56bn.pdf.

Naughton, Barry, and Kellee Tsai. 2015. *State Capitalism, Institutional Adaptation, and the Chinese Miracle*. New York: Cambridge University Press.

Norris, William J. 2016. *Chinese Economic Statecraft: Commercial Actors, Grand Strategy, and State Control*. Ithaca, NY: Cornell University Press.

Oatley, Thomas. 2011. "The Reductionist Gamble: Open Economy Politics in the Global Economy." *International Organization* 65, no. 2: 311–41.

OECD. 2008. *Reviews of Innovation Policy: China*. Paris: OECD.

———. 2009. "Enhancing Market Openness through Regulatory Reform." In *OECD Reviews of Regulatory Reform: China 2009; Defining the Boundary between the Market and the State*. Paris: OECD.

Oh, Seung-Youn. 2015. "How China Outsmarts WTO Rulings in the Wind Industry." *Asian Survey* 55, no. 6: 1116–45.

Pandya, Sonal. 2014. *Trading Spaces: Foreign Direct Investment Regulation, 1970–2000*. Political Economy of Institutions and Decisions. Cambridge: Cambridge University Press.

———. 2016. "Political Economy of Foreign Direct Investment: Globalized Production in the Twenty-First Century." *Annual Review of Political Science* 19, no. 1: 455–75.

Pearson, Margaret. 1991. *Joint Ventures in the People's Republic of China: The Control of Foreign Direct Investment under Socialism*. Princeton, NJ: Princeton University Press.

———. 2001. "The Case of China's Accession to GATT/WTO." In *The Making of Chinese Foreign and Security Policy in the Era of Reform*, ed. David M. Lampton, 337–70. Stanford, CA: Stanford University Press.

———. 2005. "The Business of Governing Business in China: Institutions and Norms of the Emerging Regulatory State." *World Politics* 57, no. 2: 296–322.

———. 2007. "Governing the Chinese Economy: Regulatory Reform in the Service of the State." *Public Administration Review* 67, no. 4: 718–30.

———. 2015. "State-Owned Business and Party-State Regulation in China's Modern Political Economy." In *State Capitalism, Institutional Adaptation, and the Chinese Miracle*, ed. Barry Naughton and Kellee S. Tsai, 27–45. New York: Cambridge University Press.

Pecht, Michael. 2006. *China's Electronics Industry: The Definitive Guide for Companies and Policy Makers with Interest in China*. Norwich, NY: William Andrew.

Pei, Minxin. 2016. *China's Crony Capitalism: The Dynamics of Regime Decay*. Cambridge, MA: Harvard University Press.

Pelc, Krzysztof J. 2011. "Why Do Some Countries Get Better WTO Accession Terms Than Others?" *International Organization* 65, no. 4: 639–72.

———. 2014. "The Politics of Precedent in International Law: A Social Network Application." *American Political Science Review* 108, no. 3: 547–64.

Perry, Elizabeth J. 2011. "From Mass Campaigns to Managed Campaigns: Constructing a 'New Socialist Countryside.'" In *Mao's Invisible Hand: The Political Foundations of Adaptive*

Governance in China, ed. Sebastian Heilmann and Elizabeth J. Perry, 30–61. Cambridge, MA: Harvard University Press.

Petersmann, Ernst-Ulrich. 1997. *The GATT/WTO Dispute Settlement System: International Law, International Organizations and Dispute Settlement*, Vol. 23. Nijhoff Law Specials. Boston: Kluwer Law International.

Pierson, Paul. 1993. "When Effect Becomes Cause: Policy Feedback and Political Change." *World Politics* 45, no. 4: 595–628.

——. 2000. "Increasing Returns, Path Dependence, and the Study of Politics." *American Political Science Review* 94, no. 2: 251–67.

Pinto, Pablo M. 2013. *Tying Hands or Exchanging Hostages: Partisan Governments, Commitment, and FDI Performance*. Cambridge: Cambridge University Press.

Qian, Yingyi, and Jinglian Wu. 2003. "China's Transition to a Market Economy: How Far across the River?" In *How Far across the River? Chinese Policy Reform at the Millennium*, ed. Nicholas Hope, Dennis Tao Yang, and Mu Yang Li, 31–63. Stanford Studies in International Economics and Development. Stanford, CA: Stanford University Press.

Qin, Julia Ya. 2003. "'WTO-Plus' Obligations and Their Implications for the World Trade Organization Legal System." *Journal of World Trade* 37, no. 3: 483–522.

——. 2007. "Trade, Investment and Beyond: The Impact of WTO Accession on China's Legal System." *China Quarterly* 191, no. 4: 720–41.

Rawski, Thomas G. 1999. "Reforming China's Economy: What Have We Learned?" *China Journal* 41 (January): 139–56.

Rithmire, Meg. 2014. "China's 'New Regionalism': Subnational Analysis in Chinese Political Economy." *World Politics* 66, no. 1: 165–94.

Rodrigues, Gilberto. 2003. "Impacts of Globalization on Strategies for Competition in Subnational Governments—the Case of Brazil." The Forum of Federations, http://www.forumfed.org/library/impacts-of-globalization-on-strategies-for-competition-in-subnational-governments-the-case-of-brazil/.

Rodrik, Dani. 1998. "Why Do More Open Economies Have Bigger Governments?" *Journal of Political Economy* 106, no. 5: 997–1032.

——. 2011. *The Globalization Paradox: Democracy and the Future of the World Economy*. New York: Norton.

Rogowski, Ronald. 1987a. "Political Cleavages and Changing Exposure to Trade." *American Political Science Review* 81, no. 4: 1121–37.

——. 1987b. "Trade and the Variety of Democratic Institutions." *International Organization* 41, no. 2: 203–23.

Rooij, Benjamin van. 2009. "The Politics of Law in China: Enforcement Campaigns in the Post-Mao PRC." SSRN Scholarly Paper ID 1368181. Social Science Research Network, https://papers.ssrn.com/sol3/papers.cfm?abstract_id=1368181.

Rosendorff, B. Peter, and Helen V. Milner. 2001. "The Optimal Design of International Trade Institutions: Uncertainty and Escape." *International Organization* 55, no. 4: 829–57.

Ruggie, John Gerard. 1982. "International Regimes, Transactions, and Change: Embedded Liberalism in the Postwar Economic Order." *International Organization* 36, no. 2: 379–415.

Saich, Anthony. 2002. "The Blind Man and the Elephant: Analysing the Local State in China." In *East Asian Capitalism: Conflicts, Growth and Crisis*, ed. Luigi Tomba, 75–99. Milano: Fondazione Giangiacomo Feltrinelli.

Salazar-Xirinachs, José Manuel, Irmgard Nübler, and Richard Kosul-Wright. 2014. *Transforming Economies: Making Industrial Policy Work for Growth, Jobs and Development*. Geneva: International Labor Organization.

Schiller, Wendy J. 1999. "Trade Politics in the American Congress: A Study of the Interaction of Political Geography and Interest Group Behavior." *Political Geography* 18, no. 7: 769–89.

Schoppa, Leonard J. 1993. "Two-Level Games and Bargaining Outcomes: Why Gaiatsu Succeeds in Japan in Some Cases but Not Others." *International Organization* 47, no. 3: 353–86.

Schrank, Andrew, and Josh Whitford. 2009. "Industrial Policy in the United States: A Neo-Polanyian Interpretation." *Politics & Society* 37, no. 4: 521–53.

Segal, Adam. 2010. *Digital Dragon: High-Technology Enterprises in China.* Ithaca, NY: Cornell University Press.

Shadlen, Kenneth C. 2005. "Exchanging Development for Market Access? Deep Integration and Industrial Policy under Multilateral and Regional-Bilateral Trade Agreements." *Review of International Political Economy* 12, no. 5: 750–75.

Shaffer, Gregory. 2015. "How the World Trade Organization Shapes Regulatory Governance." *Regulation & Governance* 9, no. 1: 1–15.

Shambaugh, David. 2008. *China's Communist Party: Atrophy and Adaptation.* Washington, DC: Woodrow Wilson Center Press.

Shih, Victor. 2008. *Factions and Finance in China: Elite Conflict and Inflation.* New York: Cambridge University Press.

Shih, Victor, Christopher Adolph, and Mingxing Liu. 2012. "Getting Ahead in the Communist Party: Explaining the Advancement of Central Committee Members in China." *American Political Science Review* 106, no. 1: 166–87.

Shirk, Susan L. 1993. *The Political Logic of Economic Reform in China.* Berkeley: University of California Press.

Simmons, Beth. 2014. "Bargaining over BITs, Arbitrating Awards: The Regime for Protection and Promotion of International Investment." *World Politics* 66, no. 1: 12–46.

Skinner, G. William, and Edwin A. Winckler. 1969. "Compliance Succession in Rural Communist China: A Cyclical Theory." In *A Sociological Reader on Complex Organizations,* ed. Amitai Etzioni, 410–38. New York: Holt, Rinehart and Winston.

Skocpol, Theda. 1992. *Protecting Soldiers and Mothers: The Political Origins of Social Policy in the United States.* Cambridge, MA: Belknap Press of Harvard University Press.

Slater, Dan, and Sofia Fenner. 2011. "State Power and Staying Power: Infrastructural Mechanisms and Authoritarian Durability." *Journal of International Affairs* 65, no. 1: 15–29.

Smith, Mitchell P. 2005. "Germany's Quest for a New EU Industrial Policy: Why It Is Failing." *German Politics* 14, no. 3: 315–31.

Song, Yuwu. 2013. *Biographical Dictionary of the People's Republic of China.* Jefferson, NC: McFarland.

State Economic System Reform Commission. 1999. "Remarks by Vice-Premier Zhu Rongji at a Discussion Session with the SESRC Work Department Representatives and Central Party School SESRC Cadres" (Zhu Rongji Fuzongli Zai Quanguo Tigaiwei Gongzuo Huiyi Bufen Daibiao he Zhongyang Dangxiao Tigaiban Xueyuan Zuotanhui Shang de Jianghua). In *Speeches from the Central Committee and State Council Leadership regarding SESRC Knowledge and System Reform Work (Zhongyang, Guowuyuan Lingdao Tongzhi dui Guojia Tigaiwei Gongzuo de Zhishi he Guanyu Tigai Gongzuo de Jianghua),* 175–83. Zhonggong Zhongyao Lishi Wenxian Ziliao Huibian. Los Angeles: Zhongwen Chubanwu Fuwu Zhongxin.

Steinberg, Richard H. 2006. "The Transformation of European Trading States." In *The State after Statism: New State Activities in the Age of Liberalization,* ed. Jonah D. Levy, 340–66. Cambridge, MA: Harvard University Press.

Steinfeld, Edward. 2012. *Playing our Game: Why China's Rise Doesn't Threaten the West.* New York: Oxford University Press.

Stiglitz, Joseph E., Justin Yifu Lin, and Célestin Monga. 2013. "The Rejuvenation of Industrial Policy." Policy Research Working Paper No. 6628, Washington, DC: World Bank, https://openknowledge.worldbank.org/handle/10986/16845.

Suttmeier, Richard P., and Xiangkui Yao. 2008. *Standards and the State: Chinese Technology Policy in an Age of Globalisation*. Singapore: World Scientific.

Suttmeier, Richard P., Xiangkui Yao, and Alex Zixiang Tan. 2009. "Standards of Power? Technology, Institutions, and Politics in the Development of China's National Standards Strategy." *Geopolitics, History and International Relations* 1, no. 1: 46–84.

Tallberg, Jonas. 2002. "Paths to Compliance: Enforcement, Management, and the European Union." *International Organization* 56, no. 3: 609–43.

Tamayo, Arturo Borja. 2001. "The New Federalism in Mexico and Foreign Economic Policy: An Alternative Two-Level Game Analysis of the Metalclad Case." *Latin American Politics and Society* 43, no. 4: 67–90.

Tan, Yeling. 2020. "Disaggregating 'China, Inc.': The Hierarchical Politics of WTO Entry." *Comparative Political Studies* 53, no. 13: 2118–52.

Thelen, Kathleen. 2004. *How Institutions Evolve: The Political Economy of Skills in Germany, Britain, the United States, and Japan*. Cambridge Studies in Comparative Politics. New York: Cambridge University Press.

———. 2012. "Varieties of Capitalism: Trajectories of Liberalization and the New Politics of Social Solidarity." *Annual Review of Political Science* 15, no. 1: 137–59.

Thielemann, Eiko. 1999. "Institutional Limits of a Europe with the Regions: EC State-Aid Control Meets German Federalism." *Journal of European Public Policy* 6, no. 3: 399–418.

Thun, Eric. 2006. *Changing Lanes in China: Foreign Direct Investment, Local Governments, and Auto Sector Development*. New York: Cambridge University Press.

Tiefer, Charles. 1998. "Free Trade Agreements and the New Federalism." *Minnesota Journal of Global Trade* 7: 45–74.

Tomz, Michael. 1997. "Do International Agreements Make Reforms more Credible? The Impact of NAFTA on Mexican Stock Prices." Paper presented at the Annual Meeting of the American Political Science Association, Washington DC.

Tsai, Kellee S. 2002. *Back-Alley Banking: Private Entrepreneurs in China*. Ithaca, NY: Cornell University Press.

Tsai, Lily L. 2007. *Accountability without Democracy: Solidary Groups and Public Goods Provision in Rural China*. Cambridge Studies in Comparative Politics. New York: Cambridge University Press.

Vogel, Steven. 2018. *Marketcraft: How Governments Make Markets Work*. New York: Oxford University Press.

Vu-Thanh, Tu-Anh. 2017. "Does WTO Accession Help Domestic Reform? The Political Economy of SOE Reform Backsliding in Vietnam." *World Trade Review* 16, no. 1: 85–110.

Wade, Robert. 2003. "What Strategies Are Viable for Developing Countries Today? The World Trade Organization and the Shrinking of 'Development Space.'" *Review of International Political Economy* 10, no. 4: 621–44.

———. 2014. "The Paradox of US Industrial Policy: The Developmental State in Disguise." In *Transforming Economies: Making Industrial Policy Work for Growth, Jobs and Development*, ed. José Manuel Salazar-Xirinachs, Irmgard Nübler, and Richard Kozul-Wright, 379–400. Geneva: International Labor Organization.

Walmsley, Terrie L., Thomas W. Hertel, and Elena Ianchovichina. 2006. "Assessing the Impact of China's WTO Accession on Investment." *Pacific Economic Review* 11, no. 3: 315–39.

Wang, Jisi, et al. 2018. "Did America Get China Wrong?" *Foreign Affairs*, July/August 2018, https://www.foreignaffairs.com/articles/china/2018-06-14/did-america-get-china-wrong.

Wang, Yong. 2000. "China's Domestic WTO Debate." *China Business Review* 27, no.1: 54–62.

———. 2002. "China's Stakes in WTO Accession: The Internal Decision-making Process." In *China's Accession to the World Trade Organization: National and International Perspectives*, ed. Heike Holbig and Robert Ash, 20–40. Abingdon, UK: Routledge.

Way, Lucan, and Steven Levitsky. 2007. "Linkage, Leverage, and the Post-Communist Divide." *East European Politics and Societies* 21, no. 1: 48–66.

Webster, Timothy. 2014. "Paper Compliance: How China Implements WTO Decisions." *Michigan Journal of International Law* 35, no. 3: 525–78.

Weingast, Barry R., Yingyi Qian, and Gabriella Montinola. 1995. "Federalism, Chinese Style: The Political Basis for Economic Success in China." *World Politics* 48, no. 1: 50–81.

Weiss, Linda. 2005. "Global Governance, National Strategies: How Industrialized States Make Room to Move under the WTO." *Review of International Political Economy* 12, no. 5: 723–49.

World Bank. 2020. "World Bank Open Data." https://data.worldbank.org/.

Wu, Jinglian. 2008. "Institutional Reflections from China's 30 Years of Economic Reform" (Zhongguo Jingji Gaige Sanshinian Licheng de Zhidu Sikao). *China Market*, no. 46: 12–14.

——. 2013. *Wu Jinglian: Voice of Reform in China.* Cambridge, MA: MIT Press.

Wu, Mark. 2016. "The 'China, Inc.' Challenge to Global Trade Governance." *Harvard International Law Journal* 57, no. 2: 261–324.

Xue, Muqiao. 2010. *Chinese Economists on Economic Reform: Collected Works of Xue Muqiao.* Oxford, UK: Routledge, 2010.

Yang, Dali. 2002. "Can the Chinese State Meet Its WTO Obligations? Government Reforms, Regulatory Capacity, and WTO Membership." *American Asian Review* 20, no. 2: 191–221.

——. 2004. *Remaking the Chinese Leviathan.* Stanford, CA: Stanford University Press.

Yang, Hongxing, and Dingxin Zhao. 2014. "Performance Legitimacy, State Autonomy and China's Economic Miracle." *Journal of Contemporary China* 24, no. 91: 1–19.

Yao, Yang, and Muyang Zhang. 2015. "Subnational Leaders and Economic Growth: Evidence from Chinese Cities." *Journal of Economic Growth* 20, no. 4: 405–36.

Yeo, Yukyung. 2009. "Remaking the Chinese State and the Nature of Economic Governance? The Early Appraisal of the 2008 'Super-Ministry' Reform." *Journal of Contemporary China* 18, no. 62: 729–43.

Zeng, Ka, and Wei Liang, eds. 2013. *China and Global Trade Governance: China's First Decade in the World Trade Organization.* New York: Routledge.

Zhang, Xiaoji. 1997. *China's Industrial Policies in the Process of Marketization.* Tokyo: Club Foundation for Global Studies.

Zhang, Xiaowen, and Xiaoling Li. 2014. "The Politics of Compliance with Adverse WTO Dispute Settlement Rulings in China." *Journal of Contemporary China* 23, no. 85: 143–60.

Zheng, Yongnian. 2004. *Globalization and State Transformation in China.* Cambridge Asia-Pacific Studies. Cambridge: Cambridge University Press.

Zhu, Boliang. 2012. "Essays on the Political Economy of Foreign Direct Investment." PhD diss., Columbia University.

Zhu, Rongji. 2011a. "Discussion in Meeting with United States Trade Representative Barshefsky and Others" (Huijian Meiguo Maoyi Daibiao Baershefusiji Deng Shi de Tanhua). In *Zhu Rongji on the Record* (Zhu Rongji Jiang Hua Shi Lu), 352–74. Beijing: Renmin Chubanshe.

——. 2011b. "On the Relationship between Building the Party and Building the Economy" (Guanyu Dang de Jianshe yu Jingji Jianshe de Guanxi). In *Zhu Rongji on the Record* (Zhu Rongji Jiang Hua Shi Lu), 149-161. Beijing: Renmin chubanshe.

——. 2015. *Zhu Rongji on the Record: The Road to Reform, 1998–2003.* Washington, DC: Brookings Institution Press.

Zhu, Yuchao. 2011. "'Performance Legitimacy' and China's Political Adaptation Strategy." *Journal of Chinese Political Science* 16, no. 2: 123–40.

Zong, Hairen. 2002a. "Before and after the Spring Festival." *Zhu Rongji in 1999*, special issue of *Chinese Law and Government* 35: 27–35.

———. 2002b. "Visit to the United States." *Zhu Rongji in 1999*, special issue of *Chinese Law and Government* 35: 36–52.

Zysman, John. 1977. *Political Strategies for Industrial Order: State, Market, and Industry in France*. Berkeley: University of California Press.

Index

accession protocol/commitments: authoritarianism in durability of, 119; in bureaucratic tensions, 157; in the domestic politics of globalization, 153; federalism in accountability to, 162–163; implementation of, 47–48, 60–62, 63–64, 109–110; in institutional change, 1–2; in the political economy of entry, 20–27, 35; in state strategies, 39–41, 54, 57; in WTO leverage, 97. *See also* rules, WTO; transitional review mechanism

accountability: in FDI policy divergence, 145; in federal systems, 159–163; in globalization, 151–152, 155; hierarchical politics in, 61–62, 63–65, 81, 82–83, 84–88; in integrating China, 9–10, 11–13, 14–15; in the political economy of entry, 29; in the rise of state capitalism, 90, 93–96, 98–99, 108–109, 114–115, 116–118, 149; in state strategies, 36–39, 46–50, 54–58, 84–86, 148, 149, A.3–A.9

activism, policy, 126–127, 136–145, 196n79

adjustment, domestic, 20, 36, 41–42, 151–152, 156, 157. *See also* reform

advancement, political: in authoritarian regimes, 164; and corruption, 187n12; economic growth in, 10, 14, 16, 186nn8–9; and globalization, 151–152, 154–155; hierarchical politics in, 62–63, 64; and integrating China, 8, 9–10, 11–13, 14–15, 16–17; in the quest for national champions, 120–121, 123, 145; in the rise of state capitalism, 94, 96–97, 118–119; in state

strategies, 36, 38–39, 40, 41–42, 45–46, 50–55, 57–58, 148

agencies: developmental, 12–13, 97–100, 106–107, 115–116, 117; industrial, 54, 91–92, 100–101; technocratic, 19, 28, 41, 42–43, 54, 96–97, 110–111. *See also* bureaucracy/bureaucratic agencies

agencies, regulatory: in integrating China, 13; in the rise of state capitalism, 89–90, 96–99, 101–103, 110, 115, 117; in state strategies, 40–41, 54, 57; in the WTO entry challenge, 19; WTO rules as leverage for, 110, 149. *See also* bureaucracy/bureaucratic agencies

agriculture, 22–23, 26, 31–32, 39, 51

Anhui province, 82

anticorruption campaign of Xi Jinping, 116–117

applications of China's case, 157–159

Asian financial crisis of 1998, 22

authoritarianism/authoritarian regimes, 14–15, 19, 29–31, 40, 92–100, 119, 148, 163–165

automotive industry/sector, 22, 82, 125–128, 131–133, 135–136, 139–145

autonomy in policymaking: in federalist systems, 160; in generalizability of the Chinese case, 157; hierarchical politics in, 74–75, 84–85; in integrating China, 8; international rules threatening, 28, 41; in the rise of state capitalism, 94–96, 112; in state strategies, 37, 39, 42–43, 48